T0239016

Indexed by SCOPUS Nanoscience and nanotechnology offer means to assemble and study superstructures, composed of nanocomponents such as nanocrystals and biomolecules, exhibiting interesting unique properties. Also, nanoscience and nanotechnology enable ways to make and explore design-based artificial structures that do not exist in nature such as metamaterials and metasurfaces. Furthermore, nanoscience and nanotechnology allow us to make and understand tightly confined quasi-zero-dimensional to two-dimensional quantum structures such as nanopalettes and graphene with unique electronic structures. For example, today by using a biomolecular linker, one can assemble crystalline nanoparticles and nanowires into complex surfaces or composite structures with new electronic and optical properties. The unique properties of these superstructures result from the chemical composition and physical arrangement of such nanocomponents (e.g., semiconductor nanocrystals, metal nanoparticles, and biomolecules). Interactions between these elements (donor and acceptor) may further enhance such properties of the resulting hybrid superstructures. One of the important mechanisms is excitonics (enabled through energy transfer of exciton-exciton coupling) and another one is plasmonics (enabled by plasmon-exciton coupling). Also, in such nanoengineered structures, the light-material interactions at the nanoscale can be modified and enhanced, giving rise to nanophotonic effects.

These emerging topics of energy transfer, plasmonics, metastructuring and the like have now reached a level of wide-scale use and popularity that they are no longer the topics of a specialist, but now span the interests of all "end-users" of the new findings in these topics including those parties in biology, medicine, materials science and engineerings. Many technical books and reports have been published on individual topics in the specialized fields, and the existing literature have been typically written in a specialized manner for those in the field of interest (e.g., for only the physicists, only the chemists, etc.). However, currently there is no brief series available, which covers these topics in a way uniting all fields of interest including physics, chemistry, material science, biology, medicine, engineering, and the others.

The proposed new series in "Nanoscience and Nanotechnology" uniquely supports this cross-sectional platform spanning all of these fields. The proposed briefs series is intended to target a diverse readership and to serve as an important reference for both the specialized and general audience. This is not possible to achieve under the series of an engineering field (for example, electrical engineering) or under the series of a technical field (for example, physics and applied physics), which would have been very intimidating for biologists, medical doctors, materials scientists, etc.

The Briefs in NANOSCIENCE AND NANOTECHNOLOGY thus offers a great potential by itself, which will be interesting both for the specialists and the non-specialists.

Mengtao Sun · Xijiao Mu

# Computational Simulation in Nanophotonics and Spectroscopy

Mengtao Sun
University of Science and Technology
Beijing
Beijing, China

Xijiao Mu
University of Science and Technology
Beijing
Beijing, China

ISSN 2191-530X          ISSN 2191-5318  (electronic)
SpringerBriefs in Applied Sciences and Technology
ISSN 2196-1670          ISSN 2196-1689  (electronic)
Nanoscience and Nanotechnology
ISBN 978-981-99-4731-7          ISBN 978-981-99-4732-4  (eBook)
https://doi.org/10.1007/978-981-99-4732-4

Jointly published with Tsinghua University Press
The print edition is not for sale in China. Customers from China please order the print book from: Tsinghua University Press.

This Springer imprint is published by the registered company Springer Nature Singapore Pte Ltd.
The registered company address is: 152 Beach Road, #21-01/04 Gateway East, Singapore 189721, Singapore

Paper in this product is recyclable.

# Acknowledgement

This work was supported by the National Natural Science Foundation of China (91436102, 11374353 and 11874084) and the Fundamental Research Funds for the Central Universities.

# Contents

# Chapter 1
# Introduction

Nanophotonics and spectroscopy have developed rapidly in recent years [1, 2]. The experimental research on nanophotonics is blooming. In addition to experimental research on the principles and applications of nanophotonics, computational simulation research on its various physical mechanisms and phenomena is equally important [3–5]. The simulation of the optical properties of molecules or crystals, such as electronic spectra (absorption and emission spectra, etc.), vibrational spectroscopy has an extraordinary guiding significance for experiments. In addition to the calculation and simulation of phenomena, the current computational simulation technology can also explain and analyze the physical mechanisms contained in these phenomena [6]. However, there are many modern computing simulation technology programs, and the usage methods and applicable scenarios are also very different. For users who are new to computational simulation, the threshold for conducting scientific research activities based on computational simulation is very high. For researchers who already have a certain foundation, it is difficult to master all kinds of software programs, strange keywords and programming languages, and a series of script auxiliary programs. Therefore, beginners need a detailed book to get started with this technology, and those who have the basics need a desk book to quickly query commands and script usage. Therefore, this book is an introductory book and a long-term desk book for graduate students and general researchers who need to be engaged in photonics and optics.

You will encounter three concepts in computing simulation: modeling, computing and simulation. Modeling is the premise of computation and simulation, and refers to the logical process of combining actual scenes with calculations, such as how to choose methods, how to set up systems, and load conditions [7]. Computation is parallel to simulation. Generally, computation is to obtain a configuration (molecular structure) or properties of a small system, such as first-principles calculation and DFT. Computation requires relatively high accuracy of results, and generally requires mathematical knowledge, so it can be called calculation. Simulation focuses on the description of "evolution", such as phase field method and molecular dynamics, and

© Tsinghua University Press 2023

M. Sun and X. Mu, *Computational Simulation in Nanophotonics and Spectroscopy*,
Nanoscience and Nanotechnology,
https://doi.org/10.1007/978-981-99-4732-4_1

is generally used to test whether the conclusions obtained from experiments can be explained by microscopic processes. There may be a phenomenological theory in simulation. The phenomenological theory is a simplification of the real scene and is used to expand the processing scale. The finite element method is also generally classified as simulation (maybe because it is often inaccurate). For example, for metal matrix composites, the simulation of mechanical properties is based on finite element, modelling is the geometric model and the constitutive relationship; the interface bonding simulation is based on molecular dynamics, and modelling is the possible orientation relationship and atomic species; the physical properties of the precipitated phase are calculated Based on first principles, modelling is the type of lattice, etc. A large number of modeling and calculations exist in materials science and engineering. Computational materials science has been widely discussed as early as around 1990, and there are many outputs. Computational materials science is not an "aspect", it is a large subject, and the gap between different methods is like ceramics and polymers. You need to be clear enough about what you want to do, what method to use, and then read the literature extensively to judge the prospects. Machine learning has been widely used in materials science. Typical examples are high-entropy alloy composition design and amorphous metals. It should be noted that the machine learning method is only a data processing method. It is good at discovering laws from data, but data is the most rare in materials science.

In materials research, why do we need rational design and computational simulation? In other words, what is the difference between material rational design and computational simulation and traditional research, and what opportunities does it bring to materials science research? [8, 9] What is the development trend of China and the core science and technology that need to be focused and prioritized? As we all know, the development of advanced materials is the basis of scientific and technological progress and the forerunner of high-tech development, and is the key to improving national competitiveness. For a long time, the field of materials science has relied on experiment and empirical model theory. Understanding the properties of different materials, adopt the model of "experience-guided experiment", optimize material performance and explore new materials along the chain of "processing-structure-property-performance". The development cycle is long and the cost is high. Shorten the development cycle of new materials and reduce. The research and development cost, and finally the realization of the "reverse" material structure, composition and processing plan from the specific performance is the ultimate goal of materials science research. The realization of this goal requires the rational design and computational simulation of materials. Understanding of materials has gone from the macro to the micro, and has experienced the scientific paradigm of "experience" and "models". For example, the law of thermodynamics is a theoretical model with broad meaning. But over time, for many scientific problems in materials, theoretical models Increasingly complex, it is impossible to analyze and solve directly. In recent decades, the rapid development of computers has allowed people to calculate and simulate the complex real world based on "theoretical models", which is a new paradigm for materials science research. As the third paradigm of materials science research, The rational design and computational simulation of materials can be regarded as the use of computers to carry out "experimental" research (theoretical Experiment). By preparing "samples" (building models), the development of advanced "experimental instruments" (theoretical methods, calculation methods and procedures, etc.), Obtain the original "experimental data" (calculation results). Compared with traditional experimental research, the rational design of materials is Computational simulation can not only use theoretical models to process, summarize and analyze computational data to obtain observable physical quantities, but also can reveal the structure of materials at the atomic level through electronic structure calculations based on first principles. The internal connection between physical properties and the exploration of microscopic mechanisms. It is this difference that allows us to predict the structure and performance of materials through computational simulation research, and establish the relationship between the structure and physical properties of the material at the atomic level, and further design for specific properties New materials. The research mode of materials has changed from the traditional "experience guided experiment" to "rational design and calculation simulation, experimental verification", shortening the research and development cycle of new materials. In addition, the fact that cannot be ignored is that computers are getting cheaper and unit calculations Capabilities are getting stronger and stronger; on the contrary, the requirements for experimental technology in the exploration of new materials are getting higher and higher, and the cost of experiments is increasing rapidly; if the rational design and computational simulation of materials can quickly provide sufficiently accurate results, at the same time, the computational cost will become more than experimental. When it is lower, the R&D cost of new materials can be reduced through the rational design and computational simulation of materials. The most challenging core problem facing the rational design and computational simulation of materials is how to quickly obtain sufficiently accurate results to guide experimental research The answer to this question is the "experimental instrument" that the Institute of Material Rational Design and Computational Simulation relies on. Therefore, the core science and technology that we need to focus on and prioritize is the development of advanced theoretical calculation methods and software.

The real material system spans a large time and space scale, and the issues involved are complex. Structure is the most basic problem in material science research and the basis of material performance. Due to the highly complex potential energy surface of materials It has long been a challenge to predict the possible structure of a material from a given chemical composition. The performance of a material is determined by the nature of the material's response to the external field (force, heat, light, electricity, magnetism, etc.) in the service environment These problems all depend on the electronic state of the material, and the description of the electronic structure of the material system often requires the calculation of the electronic structure based on first principles. Although the world has taken the lead in developing a series of quantum chemical calculation methods and software, Initially and satisfactorily solve the problem of the ground state electronic structure of simple material systems. However, the performance of real material systems generally involves electronic excited states, excited state evolution dynamics, and surface reaction dynamics. Development of electronic structure calculation methods and programs can be Describing the electronic structure of complex material systems quickly and accurately is the key to determining the efficiency and reliability of material rational design and computational simulation. On the other hand, with the improvement of experimental technology and computing power, we are in the material science research The ability to collect "big data" has greatly exceeded our ability to analyze this data. The development trend of rational design and computational simulation: collecting and integrating data generated by experiments and computational simulations in materials science research, combining basic physical principles, databases, information science, and machine learning technology to decipher the material "processing-structure-property- The hidden relationship in the performance" research chain is driven by "big data" to drive the rational design and computational simulation of new materials. Although similar "big data" driving technologies have been widely used in many fields, the unique data complexity of the material field And diversity requires the development of new big data methods and procedures to further promote the rapid, quantitative and reliable prediction of material structure, performance and dynamic evolution process, and finally realize the rational design driven by material "big data". Under the background, in 2011, the National Science and Technology Commission of the United States issued the "Material Genome Project", which combines three technologies such as high-throughput computing, experiments and special databases to collaboratively improve the level of material manufacturing and enhance the national competitiveness of the United States. At the same time, Chinese scientists have also taken active actions to meet this opportunity and challenge, and laid out China's Material Genome Project [10]. It can be said that the core of the Material Genome Project is to rationally design materials based on calculations and big data analysis. In general, materials As a new paradigm for materials science research, the rational design and computational simulation of materials will promote the rapid development of materials science research. At the same time, the rational design and computational simulation of materials, as the core foundation of material genome genetic engineering, are urgently developing independent intellectual property rights and international leadership The computing method and software platform of the company will be the key to whether we can seize this opportunity and achieve a leading position.

This book introduces the principles and applications of computational simulation problems in nanophotonics and spectroscopy. It is mainly divided into two parts: principle and program application. In terms of calculation principle, it can also be divided into quantum mechanics calculation and classical mechanics simulation. It mainly uses first-principles techniques in the framework of quantum mechanics and quantum chemistry to simulate and analyze molecular systems and solid systems and their surfaces; and use classical electromagnetic field theory to perform finite element calculations on the optical properties and phenomena of materials. Including the analysis and application of subwavelength optics and surface plasmons. While introducing the principles, theories, and physical formulas, this book will also list specific methods of calculation and simulation, such as program input files, model building methods, and subsequent analysis and drawing methods. Some batch-processing utility script programs will also be given for the convenience of users. This book aims to enable beginners to get started with computational simulation technology as soon as possible, and to get help from the book for a long time.

# Chapter 2
# Theoretical Basis of Computational Simulation

## 2.1 Semi-empirical Method

### 2.1.1 Introduction of Semi-empirical Method

Solving the Hartree–Roothaan equation after strictly calculating all integrals without any simplification is called ab initio calculation [11]. Since the quantum chemistry ab initio calculation method is time-consuming and requires large memory and disk space, people try to approximate the Roothaan equation based on quantum chemistry calculations to reduce the amount of calculation. The semi-empirical calculation of quantum chemistry simplifies the Roothaan equation on three levels: wave function, Hamilton operator and integral:

(1) Single-electron approximation: The equivalent Hamilton quantity selected without considering the two-electron interaction, such as the EHMO method.

(2) The Xn method of calculating the exchange potential energy with the statistical average model.

(3) Calculation methods based on zero differential overlap (ZDO) approximation, such as AM1, PM3, CNDO/2, INDO, NNDO, etc.

The semi-empirical approximation method estimates some of the most difficult integrals from some data on the electronic structure and ignores all three-center and four-center double electron integrals, which reduces the calculation amount of the Fock matrix from M4 to M2, which greatly simplifies Schrödinger Calculation of the solution of the equation. The main application object of the semi-empirical method is organic systems, which can be used to calculate very large systems and the calculation time is fast. The applicable system can achieve very good accuracy. The approximation introduced by the semi-empirical method greatly simplifies the necessary calculation workload, so that some more complex electronic structures can be calculated. The data obtained by this calculation has qualitative and

© Tsinghua University Press 2023
M. Sun and X. Mu, *Computational Simulation in Nanophotonics and Spectroscopy*,
Nanoscience and Nanotechnology,
https://doi.org/10.1007/978-981-99-4732-4_2

semi-quantitative characteristics. This method can only explain the nature of the molecule being studied by the accuracy of its calculation when solving some problems within its ability.

Regarding a huge system, no matter what method is combined with DFT, it can't be handled anyway (of course, a semi-empirical DFT like SCC-DFTB is another matter). If the force field is used, the accuracy is often limited, the universality is poor, and appropriate parameters are often lacking. The semi-empirical method is often the only option, although its importance has been largely ignored and forgotten by people.

Traditional semi-empirical methods, such as mainstream AM1 and PM3, and RM1 and PM6 which were later re-parameterized, are relatively poor for weak interactions. Even weak interactions with main electrostatic components such as hydrogen bonds are not good. Since diffusion correction has become popular in the ab initio calculations and DFT fields, people have combined semi-empirical methods with correction items for diffusion and other intermolecular interactions, making the semi-empirical methods have a qualitative leap in the performance of weak interactions. Including AM1-D, PM3-D, OM1/2/3-D, PM6-DH1, PM6-DH2, PM6-DH2X, PM6-DH+, PM6-D3H4 and PM7. Among them, the AM1-D and PM3-D proposed by McNamara and Hillier in 2007 are used by fewer people. The OMx-D series proposed in 2008 can only be calculated by its author Walter Thiel's MNDO2005 program. This thing is not open to everyone (the author despises this behavior), so naturally it is not popular. It is said that the weak interaction calculation of OM2-D The accuracy can reach the level of DFT-D. PM6-DH1, proposed in 2009, added diffusion and hydrogen bond correction items on the basis of PM6, and was soon replaced by the improved PM6-DH2 proposed in 2010. In 2011, PM6-DH2X added halogen bond correction for halogen-oxygen and halogen-nitrogen based on PM6-DH2. In 2012, PM6-D3H4 was proposed. In addition to the hydrogen bond and diffusion correction items, Pauli mutual exclusion correction items were also introduced. Perhaps this is the best method for correcting weak interactions based on PM6. The aforementioned several methods of modifying PM6 were all proposed by Hobza et al. In 2010, another person Korth also proposed a form of PM6 plus hydrogen bond and diffusion correction, called PM6-DH+, which is better than PM6-DH2, but Not necessarily better than PM6-D3H4, in addition, this method also has the problem that the potential energy surface is not smooth in some processes. PM7 is the latest semi-empirical method proposed by the author of PM3, PM6 and MOPAC by JJP Stewart in 2013. It has improved some of the defects of its previous generation PM6. From the statistical data, the calculation of thermodynamic properties has not improved significantly, but this time In the process of training parameters, the weak interaction system is emphatically considered, and the correction term for weak interaction is introduced, which makes the calculation of weak interaction have a great improvement compared with PM6. Because PM7 is relatively new, it is hard to say who is more accurate for weak interaction compared with PM6-D3H4 and PM6-DH+, but it should be in the middle. At present, PM7 is the best semi-empirical method overall, and PM7 is the most recommended by individuals for calculation of weak interactions in huge systems. However, even from the statistical data, PM7 and various methods of correcting PM6 are very small in the calculation error of weak interaction. Compared with the DFT-D series, there is still a gap, and the reliability gap is greater. Therefore, for DFT-D performance It is best to use DFT-D first for a system that can be calculated, but semi-empirical methods can be used as initial calculations, such as estimating the approximate energy, preliminary optimization of the structure, or doing time-consuming tasks such as dynamics and conformation search.

Semi-empirical calculations are the most successful in the description of organic chemistry, where only a few elements are widely used, and the molecules are of medium size. However, semi-empirical methods are also applied to solid and nanostructures, but with different parameterizations. Empirical research is a way of obtaining knowledge through direct and indirect observation or experience. As with empirical methods, we can distinguish the following methods: Methods restricted to $\pi$-electrons. These methods exist for calculating cyclic and linear electronic excited states of polyenes. These methods, such as Pariser–Parr–Pople Square (PPP), can estimate the $\pi$ electronic excited state very well. When the parameterization is good, in fact, the PPP method has outperformed the initial excited state calculation for many years. Methods restricted to all valence electrons. These methods can be grouped into several groups: The methods such as CNDO/2, INDO and NDDO introduced by John Pople are designed to adapt, not to experiment, but to start from scratch with minimal benchmark set results [12]. These methods are rarely used now, but the methods are usually the basis for subsequent methods. The original methods from Michael Dewar in the MADAC, AMPAC and/or SPARTAN computer programs are MINDO, MNDO, AM1, PM3, RM1, PM6 and SAM1, respectively. The goal here is to use parameters to adapt the formation heat, dipole moment, ionization potential, and geometry of the experiment. The main purpose of the method is to predict the geometry of coordination compounds, such as Sparkle/AM1, which can be used for lanthanide complexes. The main purpose of the method is to calculate the excited state to predict the electronic spectrum. These include ZINDO and SINDO, the latter being the largest method to date.

## 2.1.2 The Accuracy and Applicable Scale of the Semi-empirical Method

Semi-empirical quantum mechanics. The amount of calculation at the beginning of the calculation increases with a quarter of the basic size, and therefore, its application to macromolecules is expensive in terms of time and computer resources. Therefore, semi-empirical methods that only deal with valence electrons have been developed, in which some integrals are ignored or replaced by empirical parameters. Various semi-empirical parameterizations (MNDO, AM1, PM3, etc.) currently in use have greatly increased the molecular size that can be obtained by quantitative modeling methods and the accuracy of the results. The semi-empirical method is often suitable for the following aspects: If it is a very large molecule, only the semi-empirical method has practical calculation meaning; it is often used as the first step of high-precision calculation to obtain the initial structure of subsequent calculations, and many errors are reported. The situation is that the optimization speed is very slow due to the large basis set for optimization at the beginning, or even the optimization can not go on; in general, the semi-empirical method is suitable for simple organic molecules; qualitative information such as molecular orbital and charge can be obtained And Jianzheng mode, etc. Among the most popular semi-empirical quantum chemistry calculation methods, the most widely used methods and the methods with the highest precision and accuracy are AM1 (Austin Model 1) and PM3 (Parmetric Method 3). The results calculated by AM1, such as the molecular geometry, heat of formation, dipole moment, dissociation energy, and electron nucleophilic potential, are in good agreement with the experimental results. In this way, most chemical problems can be dealt with more accurately, and at the same time, the activation barriers of many chemical reactions and the heat generation of various compounds can be predicted more accurately. The PM3 method is a development of the AM1 method using different parameters. The parameters selected in the PM3 method are obtained by comparing the various properties of the calculated molecules with the experimental data, so the accuracy is high. Generally speaking, the PM3 method uses less data than the AM1 method when dealing with non-bonded interaction energies (such as van der Waals forces, etc.). The PM3 method is widely used in the calculation of organic molecules and can also be used for most of the main group elements that have been parameterized.

### 2.1.3  Mainstream Software

Hückel method

The Hückel method or Hückel molecular orbital method (HMO) proposed by Erich Hückel in 1930 is a very simple linear combination of atomic orbital molecular orbital (LCAO MO) in the conjugated hydrocarbon system to determine the molecular orbital energy of $\pi$ electrons, such as ethylene and benzene. This is the theoretical basis of Hückel's rule. It was subsequently extended to conjugated molecules, such as pyridine, pyrrole and furan, which contain atoms other than carbon, referred to herein as heteroatoms. The extended Hückel method developed by Roald Hoffmann is computational and three-dimensional for testing Woodward De Hoffman rule.

Extended Hückel method

The extended Hückel method is a semi-empirical quantum chemical method developed by Roald Hoffmann since 1963. It is based on the Hückel method, but the original Hückel method only considers $\pi$ orbitals, and the extended method also includes $\sigma$ orbitals. The extended Hückel method can be used to determine molecular orbitals, but it has not been very successful in determining the structural geometry of organic molecules. However, it can determine the relative energy of different geometric configurations. It involves calculating electronic interactions in a fairly simple way, where electron–electron repulsion is not explicitly included, and the total energy is simply the sum of the terms for each electron in the molecule. Since the diagonal Hamiltonian matrix elements of Wolfsberg and Helmholz are given by approximation, they associate them with diagonal elements and overlapping matrix elements. K is the Wolfsberg–Helmholtz constant, usually 1.75. In the extended Hückel method, only valence electrons are considered; the core electron energy and function should remain more or less constant between atoms of the same type. This method uses a series of parameterized energies calculated from atomic ionization potentials or theoretical methods to fill the diagonals of the Fock matrix. After filling the non-diagonal elements and diagonalizing the obtained Fock matrix, the energy (eigenvalue) and wave function (eigenvector) of the valence orbit are found. In many theoretical studies, the extended Hückel molecular orbital is usually used as a preliminary step to determine the molecular orbital through more complex methods such as the CNDO/2 method and ab initio quantum chemistry methods. Since the extended Hückel basis set is fixed, the wave function of the single particle calculation must be projected to the accurately calculated reference set. The new basic track is usually adjusted to the old track by the least square method. Only the valence electron wave function can be found by this method, and the remaining reference set must be orthogonally normalized with the calculated orbit, and then the nucleon electron function is selected with less energy. This leads to a more accurate determination of the structure and electronic properties, or, in the case of the ab initio method, faster convergence. Robert Hoffmann (Roald Hoffmann) first used this method, and Robert Burns Woodward (Robert Burns Woodward) clarified the rules of the reaction mechanism (Woodward Hoffmann's rules). He used the molecular orbital diagram of the extended Hückel theory to calculate the orbital interactions in these cycloaddition reactions. Hoffmann and William Lipscomb used a very similar method in the previous study of borohydrides. The diagonal Hamiltonian matrix elements are proportional to the overlap integral. Since the difference in electronegativity between boron and hydrogen is small, this simplification of Wolfsberg and Helmholz approximation is reasonable for borohydrides. This method is less effective for molecules containing atoms with very different electronegativity. To overcome this weakness, several groups have proposed an iterative scheme that relies on atomic charges. One method that is still widely used in inorganic and organometallic chemistry is the Fenske–Hall method. The program for the extended Hückel method is YAHHOP, which stands for "another extended Hückel molecular orbital package".

## CNDO/2

CNDO is an abbreviation for completely ignoring differences and overlaps, one of the first semi-empirical methods of quantum chemistry. It uses two approximations: kernel approximation-only explicit inclusion of foreign valence electrons and difference overlap CNDO/2 is the main version of CNDO. This method was first introduced by John Pople and his colleagues. The earlier method extended the Hückel method, which explicitly ignored the term electron–electron repulsion. It is a method of calculating electron energy and molecular orbitals. CNDO/1 and CNDO/2 were developed by explicitly including electron–electron repulsion, but ignoring many of them, approximating some of them, and fitting other substances to spectroscopic experimental data. Quantum mechanics provides equations based on the Hartree–Fock method and the Roothaan equation of CNDO for simulating atoms and their positions. These equations are solved iteratively to the point where the results do not change significantly between two iterations. It is worth noting that CNDO does not involve knowledge of chemical bonds, but uses knowledge of quantum wave functions. CNDO can be used for closed shell molecules, where electrons are completely paired in molecular orbitals and open shell molecules, and these molecules are free radicals with unpaired electrons. It is also used for calculations of solid and nanostructures.

Austin Model 1

Austin Model 1 or AM1 is a semi-empirical method for quantum calculation of molecular electronic structure in computational chemistry. It is based on the approximate neglect of the differential diatomic overlap integral. Specifically, it is a generalization that the differential diatomic overlap approximate correction ignores [1]. The related methods are PM3 and the older MINDO. AM1 was developed by Michael Dewar and colleagues and published in 1985. AM1 is an attempt to improve the MNDO model by reducing the repulsion of close atoms. The nucleus term in the MNDO equation is modified by adding eccentric attraction and repulsion Gaussian functions. The complexity of the parameterization problem increases as the number of parameters per atom increases from 7 in MNDO to 13–16 per atom in AM1. The result of AM1 calculation is sometimes used as the starting point for the parameterization of the force field in molecular modeling. AM1 is implemented in MOPAC, AMPAC, Gauss, CP2K, GAMESS (US), PC GAMESS, GAMESS (UK) and SPARTAN programs. The extension of AM1 is SemiChem Austin Model 1 (SAM1), which is implemented in the AMPAC program and treats the d orbital explicitly. The AM1 calculation model of the lanthanide complex called Sparkle/AM1 was also introduced and implemented in MOPAC2007. AM1 was recently re-parameterized, resulting in a new RM1 or Recife Model 1, which can be used in MOPAC2007, SPARTAN software, Hyperchem, etc. The extension of AM1 is AM1*, which can be used in VAMP software.

PM3

PM3 or parametric model 3 is a semi-empirical method for quantum calculation of molecular electronic structure in computational chemistry. It is based on the neglect of differential diatomic superposition approximation. The PM3 method uses the same formalism and equations as the AM1 method. The only difference is: (1) PM3 uses two Gaussian functions as the core rejection function instead of the variable number used by AM1 (one to four Gaussian for each element); (2) The values of the parameters are different. The other difference lies in the philosophy and method used in the parameterization process: AM1 obtains some parameter values from measured spectra, and PM3 treats them as optimizable values. This method was developed by

JJP Stewart and first published in 1989. It is implemented as Gauss, CP2K, GAMESS (US), GAMESS (UK), PC GAMESS, Chem3D in the MOPAC program (of which the older version is in the public domain) and related RM1, AM1, MNDO and MINDO methods and several other programs, AMPAC, ArgusLab, BOSS and SPARTAN. The original PM3 publication includes the parameters of the following elements: H, C, N, O, F, Al, Si, P, S, Cl, Br, and I. The PM3 implementation in the SPARTAN program includes PM3tm, which includes Ca, Ti, V, Cr, Mn, Fe, Co, Ni, Cu, Zn, Zr, Mo, Tc, Ru, Rh, Pd, Hf, Ta, W, Re, Os, Ir, Pt and Gd. Many other elements, mainly metals, have been parameterized in subsequent work. A lanthanide complex PM3 calculation model called Sparkle/PM3 is also introduced.

MOPAC

MOPAC is a computer program commonly used in computational chemistry. It is designed to implement semi-empirical quantum chemical algorithms, and it runs on Windows, Mac and Linux. MOPAC2016 is the current version. MOPAC2016 can use PM7, PM6, PM3, AM1, MNDO and RM1 to calculate small molecules and enzymes. The sparkle model (for lanthanide chemistry) is also available. The program is available in Windows, Linux and Macintosh. Academic users can use the program for free, while government and commercial users must purchase the software. MOPAC was mainly written by Michael Dewar's research team at the University of Texas at Austin. Its name comes from Molecular Orbital PACkage, which is also a pun on the Mopac highway that runs around Austin.

## 2.2 Hartree Fock (HF) Method

The Hartree–Fokker equation, also known as the HF equation, is an equation that applies the variational method to calculate the wave function of a multi-electron system (English: Many-body problem). It is the best in quantum physics, condensed matter physics, and quantum chemistry. One of the important equations. The HF equation is a single-electron eigen equation in form, and the eigenstate obtained is a single-electron wave function, which is a molecular orbital. The numerical calculation method with the HF equation as the core is called the "Hartree–Fock method" (Hartree–Fock method). All quantum chemical calculation methods based on molecular orbital theory are based on the HF method. In view of the wide application of molecular orbital theory in modern quantum chemistry, the HF equation is regarded as the cornerstone of modern quantum chemistry. In 1927, physicists Walter Heitler and Fritz London completed the quantum mechanical calculations of hydrogen molecules, which opened the era of quantum chemistry. Since then, people have tried to use quantum mechanics theory to explain the structure of chemical substances and chemical phenomena. In order to solve the problem of approximately solving the Schrödinger equation of the multi-electron system, quantum chemist Douglas Hartree proposed the Hartree hypothesis in 1928, treating each electron as a particle moving in the average potential field formed by all other electrons. According to his hypothesis, Hartley decomposes the system electron Hamiltonian into a simple algebraic sum of several one-electron Hamiltonian operators. Each one-electron Hamiltonian contains only the coordinates of one electron, so the system has multiple electrons. The wave function can be expressed as the simple product of the single electron wave function, which is the Hartley equation. But because Hartree did not consider the antisymmetric requirement of the electronic wave function, his Hartree equation was actually very unsuccessful.

In 1930, Hartley's students Vladimir Fokker and John Slater respectively proposed the self-consistent field iteration equation considering Pauli's principle and the wave function of the single-determinant multi-electron system. This is Hart today. Lee-Fokker equation. However, due to computational difficulties, the HF equation was silent for 20 years after its birth. In 1950, the quantum chemist Clemens Rotham thought of using a linear combination of atomic orbitals to approximate the expansion of molecular orbitals, and obtained Rotham's equation for closed shell structure.

In 1953, R. Pariser, R. Parr of the United States and John Popper of the United Kingdom spent two years using a hand-cranked calculator to independently realize the RHF self-consistent field calculation of nitrogen molecules. This is the first time for mankind. The quantum mechanical explanation of the chemical structure obtained by solving the HF equation is also the first time that the quantum chemical calculation method has been actually completed. After the first success, along with the rapid development of computer technology, the HF equation and quantum chemistry have achieved considerable development. On the basis of the HF equation, people have developed advanced quantum chemistry calculation methods, which have further improved the calculation accuracy. The simplification and parameterization of the electronic integration of the HF equation has greatly reduced the amount of quantum chemistry calculations, making it possible to calculate medium-sized molecules with more than 1,000 atoms.

The Hartree–Fokker equation originated from the variational processing of the electronic wave function of the multi-electron system. Under the Born–Oppenheimer approximation, the electron motion and energy of a multi-electron system can be separated from the motion and energy of the nucleus. In this way, the electronic energy of the system can be calculated using the electronic Hamiltonian and the multi-electron wave function. The expression of its energy is:

$$E_0 = \langle \Psi_0 | H_{\text{ele}} | \Psi_0 \rangle \tag{2.1}$$

where $E_0$ represents the ground state electron energy of the system, $H_{\text{ele}}$ represents the electronic Hamiltonian of the system

$$H_{ele} = \sum_i^N \frac{-1}{2} \nabla_i^2 - \sum_{i=1}^N \sum_{a=1}^M \frac{Z_A}{R_{iA}} + \sum_{i=1}^N \sum_{j>i}^N \frac{1}{r_{ij}} \tag{2.2}$$

According to the mode of action, $H_{\text{ele}}$ can be decomposed into two parts: $H_{ele} = O_1 + O_2$. Where $O_1$ is a single electron operator, which Describe the kinetic energy of a single electron and the attractive potential energy of the nucleus and the $O_2$ is the dual-electron integral.

$$O_1 = \sum_i^N \frac{-1}{2}\nabla_i^2 - \sum_{i=1}^N \sum_{a=1}^M \frac{Z_A}{R_{iA}} = \sum_i^N - \left(\frac{1}{2}\nabla_i^2 + \sum_{a=1}^M \frac{Z_A}{R_{iA}}\right) = \sum_i^N h_i \quad (2.3)$$

$$O_2 = \sum_{i=1}^N \sum_{j>i}^N \frac{1}{r_{ij}} \quad (2.4)$$

$\Psi_0$ stands for the ground state multi-electron wave function, which is the Slater determinant formed by the single-electron molecular orbital wave function of the system as the basis function.

$$\Psi = \frac{1}{\sqrt{N!}} \begin{vmatrix} \chi_1(1) & \chi_2(1) & \cdots & \chi_N(1) \\ \chi_1(2) & \chi_2(2) & \cdots & \chi_N(2) \\ \vdots & \vdots & \ddots & \vdots \\ \chi_1(N) & \chi_2(N) & \cdots & \chi_N(N) \end{vmatrix} = |\chi_1\chi_2\cdots\chi_N\rangle \quad (2.5)$$

The molecular orbitals constructing $\Psi_0$ are orthogonal to each other, that is, the constraints are

$$\langle \chi_a \mid \chi_b \rangle = \delta_{ab} \quad (2.6)$$

Although the Hartree–Fokker equation has a simple eigen equation form, the Coulomb operator and the commutative operator in the Fokker operator contain all $\chi_a$, so the actual form of the equation is very complicated, and an accurate analytical solution cannot be obtained. It can only be solved by iterative method, which is the self-consistent field method in quantum chemistry.

In actual operation, people will first give basis sets and convert eigen equations into matrix equations. This transformation allows the molecular orbital to be expressed as $\chi_a(x) = \sum_i C_{ai}\phi_i(x)$, you can use a varying K-dimensional vector $\mathbb{C}_{\supset}$ represents the molecular orbital, K is the number of basis functions in the basis set. In the same way, the Fokker operator can be transformed into a Fokker matrix. However, since the basis functions are not necessarily orthogonal, there are overlapping integrals

$$S_{ij} = \int dx \phi_i^*(x)\phi_j(x) \quad (2.7)$$

Finally, the form of the HF equation is transformed into a generalized eigen equation

$$\mathbb{F}\mathbb{C} = \varepsilon\mathbb{S}\mathbb{C} \quad (2.8)$$

This equation is called the Hartley–Roothaan equation, or Roothaan equation. In this way, the relatively complex eigen equations are transformed into matrix eigen equations that can be solved only by simple algebraic calculations, while the complex integrals in the original equations are completed at one time in the above conversion process.

## 2.3  Density Functional Theory

Density functional theory (DFT) is a rigorous theory for solving single electron problems. It is a flash of human wisdom and a great contribution at the Nobel Prize level. It has become a powerful tool in many fields such as physics, chemistry, and materials. A basic theory that all colleagues who study computational condensed matter physics/computational materials science/computational chemistry must learn.

### 2.3.1  Difficulties in Calculation of Actual Materials

In the early twentieth century, when people studied the microcosm, they discovered that they were very different from the macroscopic laws, and the physics framework established by classical mechanics was no longer applicable. Through the exploration of a group of outstanding physicists, people have established quantum mechanics to describe the objective laws of the microscopic world. Quantum mechanics tells us that the motion of microscopic particles follows the Schrodinger equation, and the Schrodinger equation can be written in its basic form and the form of the stationary Schrodinger equation. Stationary state means that the energy has a certain value, and the Hamiltonian is an operator that describes the total energy of the system. As the solution of Schrödinger's equation, the wave function of a system contains all the information of a system in a certain state, which provides a theoretical possibility for simulating any system, so that Dirac once said: "Most physical problems and All chemistry problems have been solved in principle, and the remaining problem is to solve the Schrodinger equation." However, it is very difficult to solve the wave function. In fact, only a few systems have analytical solutions, such as hydrogen-like systems. The system that can be solved analytically benefits from the absence of interaction terms (this will be the focus of our later description, you can remember), and can be solved by using common mathematical and physical methods such as the separation of variables method (for specific solutions, please refer to any A textbook of quantum mechanics). However, for actual material calculations, it is faced with interacting multi-particle systems, and it is impossible to use techniques like hydrogen-like systems to find analytical solutions; even numerical solutions are separated by humans between ideal and reality. Gaps in computing power. So Dirac added: "The difficulty is that the equations that apply these laws are too complicated to solve."

## 2.3.2 Hohenberg–Kohn Theorem

For density functional theory, we must first talk about the Hohenberg–Kohn theorem which supports its theory.

**Theorem 2.1** *The ground state energy of the identical Fermi system without spin is the only functional of the particle number density $\rho(r)$.*

**Theorem 2.2** *The energy functional $E(\rho)$ takes the minimum value of the correct particle number density function $\rho(r)$ under the condition of constant number of particles, and is equal to the ground state energy.*

Theorem 2.1 shows that the particle number density is the basic variable to determine the physical properties of the ground state of a multi-particle system. All ground state physical properties of multi-particle systems. Density is a function of coordinates, and energy is a function of density, and a function of function is a functional. Theorem 2.2 shows that if the ground state particle number density function is obtained, the minimum value of the energy functional can be determined. This minimum value is equal to the energy of the ground state. This is actually the variational principle for the density function. Special note here: As mentioned above, the Hatree–Fock method is a step by step simplification from multi-particle system-multi-electron system-single-electron system, and found that the only way to solve it is to solve the single-electron problem without interaction. Here, the Hohenberg–Kohn theorem of density functional theory mentions the "particle number density", which may be misleading to students who are new to it. The author believes that the "particle" here, as stated in Theorem 2.1, refers to "identical fermions", which is different from the "nucleus and electrons" we mentioned earlier. In my opinion, everyone directly thinks it is "charge density". Kohn and Sham proposed to replace the system-wide, interactive functionals with separate, non-interaction functionals, and then use a term called commutative correlation functional $E_{xc}(\rho)$ to indicate. In other words, replace the real kinetic energy $T(\rho)$ and the potential energy $V(\rho)$ with the kinetic energy $T_s(\rho)$ and potential energy $V_H(\rho)$ that are not interacting but easy to solve and calculate. This is the core idea of density functional theory,

$$E[\rho] = T_s[\rho] + U_H[\rho] + V[\rho] + E_{xc}[\rho] \tag{2.9}$$

In other words, for the density functional theory, the energy functional is accurately represented, because the exchange correlation functional term contains all errors and unknown effects.

After completing the separation of non-interaction terms and interaction terms, according to the core idea of solving Schrodinger equation of multi-particle system provided by the Hatree–Fock method, we should find a solvable non-interaction single electron problem. Therefore, we need to start looking for single electron images. According to the definition, the expressions of charge density, single electron kinetic energy and single electron action term are given:

$$\rho(r) = \sum_i |\varphi_i(r)|^2$$
$$T_s[\rho] = \sum_i \int dr \varphi_i^*(r) \left(-\nabla^2\right) \varphi_i(r) \qquad (2.10)$$
$$U_H[\rho] = \frac{1}{2} \iint dr dr' \frac{\rho(r)\rho(r')}{|r-r'|}$$

Here, the single-electron kinetic energy term is well understood, and I want to talk about this single-electron potential energy term. Some people call this term "Hatree potential energy" because it looks a lot like the expression of electronic Coulomb interaction, and is similar to the potential energy term in the Hatree–Fock method mentioned earlier.

## 2.3.3 Exchange Correlation Functional

Through approximation, an easily calculated, non-interactive but inaccurate result is separated, and then all errors are separately collected into the exchange correlation functional and left for analysis. Therefore, the choice of the exchange correlation functional term is very important, because it will directly determine the accuracy of the calculation result Of course, the density functional theory should still be viewed in a critical light: it is not omnipotent. From a practical perspective, although density functional theory has become one of the basic tools in the research fields of physics, chemistry, materials, etc., with sufficiently high calculation accuracy, it cannot actually find the exact solution of the Schrödinger equation because of the exchange The precise form of the correlation functional can never be given accurately, which makes the calculated result and the actual result inevitably have errors. In addition, in the following important situations, density functional theory cannot give a good physical exact solution:

Calculation of electronic excited state. Although density functional theory can certainly be used to predict the properties of electronic excited states, strictly speaking, the accuracy of this prediction is not guaranteed in theory, because the Hohenberg-Kohn theorem is for the ground state;

Band gap calculation for semiconductor and insulator materials. Using the existing functional standards, the accuracy of the band gap calculated by the density functional theory is limited, and there is a common error greater than 1 eV compared with the experimental data. How to use density functional theory to accurately handle this problem is still an active research field;

Calculation of van der Waals forces between atoms and molecules. Density functional theory can hardly give accurate results in this situation. One explanation is that van der Waals force is a direct result of long-range electronic correlation effects. To solve this problem, more advanced wave function methods must be used instead of density functional theory.

Calculation of a large number of atoms. This is actually a problem of calculation cost. At present, the calculation amount of density functional theory is usually around dozens of atoms. From a physical point of view, even a drop of water contains atoms of an ultra-high order of magnitude. If you want to solve all atoms directly "simple and rude", you need to rely on computer technology or program efficiency to have very huge progress and breakthroughs. It's hard to expect. Therefore, how to link the calculation results of a very small number of atoms with real materials with physical relevance is also a problem that researchers who use density functional theory must understand.

But in any case, the significance of density functional theory is remarkable and remarkable: it tells people that the Schrödinger equation can be solved by finding the charge density function of three-dimensional variables, instead of using the wave function containing 3N-dimensional variables, which greatly simplifies the calculation. Which makes it possible to calculate and solve. It is a flash of human wisdom and a Nobel Prize-level contribution, for which Kohn won the 1998 Nobel Prize in Chemistry.

Therefore, for the understanding of density functional theory, I think it should be a combination of the following three abilities: the need to understand theoretical knowledge, which lays a solid foundation for subsequent research; the need to understand technical details, through familiarization with programming and various parameters, The computer platform has become a powerful tool for scientific research; it is necessary to understand the actual combat, test what has been learned through practice, and explore the true objective laws.

## 2.3.4 Selection of Functional

If you want to analyze the characteristics of various types of functionals in detail, you have to talk about it in a long way. A few words are not clear at all. As for the recommended literature, there is really nothing and suitable for recommendation. Although there are many reviews of functionals and many monographs, there are few comprehensive reviews, and most of the people who write such articles have studied functionals and have put forward functionals. The viewpoints and content selection are always biased. Being unobjectionable, one person says one thing. In addition, new functionals continue to appear, and older articles will be out of touch with the mainstream situation, and even mislead readers. Therefore, if you really want to be very fluent in the choice of functionals, you must carefully read DFT monographs to lay a solid foundation, and pay more attention to mainstream literature, and gradually form your own understanding. This is like testing data. One or several measurements may have a large error, but if there are too many tests, the average value will be very close to the actual value.

Since many beginners lack some common sense, here is the basic common sense. The order of calculating the accuracy of energy-related problems is usually: CCSD(T) > double-hybrid functional > appropriately selected ordinary functional > MP2 > HF > semi-Empirical method.

Below is a list of functional selection suggestions for various situations, combined with the author's calculation experience and a large number of test articles.

This article will be constantly updated to keep pace with the times and reflect the author's latest views. Only quantum chemical calculations are involved here, not first-principles calculations. The "organic system" mentioned in this article refers more precisely to a system composed of main group atoms. Therefore, the functional calculation suitable for organic calculation is also suitable for systems containing alkali and alkaline earth metals.

When the double-hybrid functional can be used and better calculation accuracy is required, usually the accuracy can be improved by one level than the ordinary functional, and it is a block away from MP2, and the time-consuming is only slightly higher than MP2. The accuracy of the double-hybrid functional is inconclusive, and it has a great relationship with the system. The conclusions of different test articles are very different. Even if it is not a weak interaction, DFT-D3 (BJ) should be added to improve the overall performance. Functionals worth considering: ·PWPB95-D3(BJ): High robustness, better performance, recommended priority. ORCA and other support, Gaussian cannot be used ·B2PLYP-D3 (BJ): the oldest in history, the accuracy is mediocre in the double hybrid functional, but it is the most widely supported by the program, and the robustness is not bad ·DSD-PBEP86-D3(BJ) and DSD-BLYP-D3(BJ): They perform very well in some test articles, but there are also many surprises, so they are risky. G16 has built-in DSD-PBEP86-D3(BJ), just write DSDP-BEP86 directly ·WB97X-2: The accuracy is the top in the double hybrid functional, and it is also very robust, but only supported by GAMESS-US and Q-Chem For the case where the double hybrid functional cannot be used, the choices are as follows: ·Calculate the excitation energy of the vertical local valence layer: PBE0 (the error is about 0.2–0.3 eV, and the average is better than B3LYP) ·Calculate the excitation energy, charge transfer, higher order valence layer, Rydberg vertical excitation of large conjugate systems: wB97XD, CAM-B3LYP ·Calculate the triplet excitation energy: M06-2X ·Calculate various excitation energies, using only one functional: wB97XD, M06-2X. If you find that the result is not good, you can try PBE0, PBE38, CAM-B3LYP. PS: Using ORCA to do TDDFT based on double hybrid functional has much better accuracy and universality ·Calculate the polarization rate: PBE0 or BHandHLYP (the latter is recommended for large conjugate systems). DSD-PBEPBE-D3 has the best polarization rate in dual hybrids, second only to CCSD. ·Calculate the hyperpolarizability: wB97 or LC-$\tau$HCTH. BHandHLYP and CAM-B3LYP can also be considered. Note: To calculate the (super) polarizability, the most ideal is to use the LC-wPBE adjusted by w, but it needs extra cost to optimize the w parameter ·Calculate NMR: KT1 (Dalton support), Gaussian support can use B97-2 (see JCTC, 10,572), but it is not as good as MP2. For carbon and hydrogen, the best practice is to use B3LYP combined with the scaling method. ·Calculate molecular vibration (both resonant and non-resonant): B3LYP ·Organic molecular structure optimization: PBE0 (DFT-D3 (BJ) correction should be added when weak interaction is obvious), M06-2X. B3LYP performance is also OK ·Calculate HOMO-LUMO gap: HSE06, B3PW91 (on the whole, the latter is better, and the hybrid parameters of HSE06 are somewhat dependent on the system) ·Calculate the thermodynamic data of the organic system (reaction energy, configuration energy difference, energy barrier, etc.): M06-2X (plus DFT-D3 correction is often better) ·Compute multi-reference system with strong features: SCAN, M06L, MN15L ·Calculate transition metal complexes: SCAN, TPSSh, BP86, B97-1, PBE0, MN15L. B3LYP can also be used, but the performance is average. It is best to add D3 (BJ) correction ·Calculate the metal organic reaction of the closed shell layer: PBE0-D3(BJ). If you can only use pure functionals, use TPSS-D3 (BJ) ·Calculate the chemical bonds between transition metals (including transition metal cluster systems): PBE, BP86, TPSS ·Calculate various weak interactions: M06-2X (more robust after adding D3), wB97X-D3. Using B3LYP-D3 (BJ) is cheaper but also less accurate. In pursuit of faster speed, use BLYP-D3 (BJ) (combining RI in ORCA is fast) ·Calculate halogen bond: M06-2X (PS: SCS(MI)MP2, a variant of MP2, performs very well) ·B3LYP is used to calculate carbon clusters, and TPSSh is used for boron clusters (PBE0 is also very commonly used, but B3LYP is not suitable) ·Calculate the energy difference of the complex with different spin multiplicity: B3LYP* (see JCP,117,4729) ·Calculate the vertical electron affinity and Fundamental gap of the organic system: wB97X (functional controlled by w is better) ·Calculate the vertical ionization energy of organic system: M06-2X ·Calculate dipole moment, electron density distribution: PBE0

B3LYP-D3 (BJ) is the first choice in other situations or ambiguities. If you find that the results are not satisfactory, you can try M06-2X, wB97XD, MN15. Of course, if you can accept a larger amount of calculation, it is best to try the double hybrid functional. The B3LYP more than 20 years has passed from 1994 to now, is still the most used functional in the quantitative world, and it is also the default functional by most people, including the author. Although it is largely because everyone is used to it, the key is that B3LYP as a whole is very durable and robust. Although there are currently hundreds of functionals, many of them can kill B3LYP in their own areas of expertise, but there are very few functionals whose comprehensive performance can exceed B3LYP, so the popularity, popularity and B3LYP are simply incomparable. B3LYP has two key weaknesses (1): Van der Waals' effect is completely impossible to describe, but it can be completely solved with DFT-D3 (BJ) correction. This correction also makes other aspects of B3LYP such as thermodynamic data calculation accuracy have a small improvement (2): The calculation of charge transfer (CT) excitation and Rydberg excitation is bad, but the variant CAM-B3LYP also largely solves this problem (but note that CAM-B3LYP is bad for other occasions). In addition, B3LYP also has some other relatively minor problems, such as the system with strong static correlation and the unsatisfactory description of binuclear metal bonds (a common problem of hybrid functionals), and the overestimation of high spin state stability for some transition metal complexes (Adjusting the HF content from 20 to 15% can be solved, some chemical reactions are not calculated properly (such as F+H2->HF+H does not predict the potential barrier )and many more. However, infinite functionals (unless it is the legendary holy grail functional, even if it is found, it will not move at all), you cannot demand that B3LYP is perfect, or that B3LYP has roughly reached an optimal balance. If it is to improve the nature of A If the parameters are optimized for accuracy, the calculation accuracy of the B nature will decrease. In short, in addition to van der Waals-led weak interaction and CT/Rydberg excitation, which are two issues that must not be used with B3LYP, a new problem is encountered. If you don't know what functional is suitable and you are too lazy to search the literature to find the most suitable The functional, so big, you can use B3LYP to calculate it first.

There are two potential substitutes for the "quasi-universal" functional of B3LYP, one is M06-2X and the other is wB97XD, both of which were born in 2008, and mainstream quantitative programs such as Gaussian are also supported. Both of these functionals have the style of B3LYP, and they are good overall, and not only do they lack the two key weaknesses of B3LYP, but these weaknesses have become their strengths. M06-2X introduces a weak interaction system when fitting parameters to make the description of the weak interaction quite good, and because the HF component is as high as 54%, it makes the calculation of CT and Rydberg excitation good. wB97XD uses DFT-D2 diffusion correction to make the calculation accuracy of weak interactions good, and introduces long-range correction to make CT and Rydberg excitations also work well. Both of these functionals have a common problem, that is, the calculation speed is slow, much slower than B3LYP (however, B3LYP can be considered mobile, and these two functionals can certainly be considered mobile). Moreover, Minnesota series functionals such as M06-2X have much higher requirements on the accuracy of DFT integration grid points than ordinary functionals. Sometimes the geometric optimization and SCF are not easy to converge due to insufficient grid points, and wrinkles appear on the potential energy surface. It can be solved by improving the accuracy of the integration grid (such as int=ultrafine in Gaussian09. Int=superfine is required in rare cases), but it will significantly increase the time-consuming. In addition, M06-2X is parameterized for the main group, which is very inappropriate for the energy calculation of transition metal systems. If you have to use the Minnesota series of functionals, you should use the same M06L, which is more suitable for transition metals. Like M06-2X, wB97XD is not suitable for energy calculation of transition metal system. Since the HF components of wB97XD and M06-2X are much higher than B3LYP, the description of static correlation is much worse, and the HOMO-LUMO gap will be overestimated, and the excitation energy of the singlet local valence layer will also be biased. High tendency. In addition, according to the test of JCTC, 6, 2115, M06-2X calculates the non-resonant frequency very badly, and wB97X is not good (so its sister wB97XD should not be good in this respect). Due to these reasons, especially the slow calculation speed, they have not been able to replace B3LYP as the default functionals, and can only be used when there is a definite need. A distinct advantage of M06-2X is that the calculation accuracy of the organic system reaction energy, isomerization energy, potential barrier and other thermodynamic quantities is quite good, sometimes even close to the accuracy of the double hybrid functional, which is also related to the fitting parameters. The introduction of such issues is directly related. In terms of thermochemical properties and structural optimization of organic systems, M06-2X is currently almost the best in non-dual hybrid functionals.

After wB97XD was proposed, one of the authors proposed wB97X-D3, changing the diffusion correction from DFT-D2 to DFT-D3, and also changed some other parameters. The accuracy of wB97X-D3 is better than that of wB97XD, including thermochemistry, weak interaction, etc. As long as the program can support wB97X-D3 (ORCA support, but Gaussian does not), then the aforementioned wB97XD is recommended to be changed to wB97X- D3.

After the M06 series of Minnesota series of functionals were proposed, although many new Minnesota series of functionals were proposed, there was no substantial improvement in accuracy for molecular systems. It will soon fall into a bottleneck), and some are obviously not as good as the M06 series. The MN15 and MN15L proposed in 2016 can be a little noticed, and they have been supported since Gaussian16. The purpose of MN15 is to describe the transition metal and the main group in a relatively balanced manner, but it is not particularly good at any aspect. The main group is obviously not as good as M06-2X, and the transition metal is not as good as M06L, so it is actually not very useful. The intention of MN15L is to replace M06L, but there is no substantial progress.

In fact, the wB97M-V proposed in 2016 basically has the highest average accuracy of all non-double hybrid functionals (except for transition metal systems), but only a few programs such as ORCA support it, while the most mainstream Gaussian does not support it. So the previous article didn't mention it much. ORCA is actually very easy to use. If you are willing to use it, then in the category of ordinary functionals, it is recommended that you use wB97M-V when calculating the energy of an organic system, although its time-consuming is significantly higher than that of B3LYP. However, if you use RIJCOSX technology to accelerate in ORCA, the calculation time of wB97M-V in ORCA is lower than the calculation time of B3LYP in Gaussian. Note that ORCA currently does not support the analytical gradient of wB97M-V. If you need to optimize the architecture and do vibration analysis, then it is recommended to use B3LYP-D3 (BJ), which is fast and results in good results.

## 2.3.5  How to Change the Functional

Quantum chemistry researchers often use an A functional to calculate one problem, and switch to a B functional to calculate another problem. It must be understood that if you want to switch functionals, there must be very legitimate reasons. There are generally three valid reasons: (1) A functional and B functional are applicable to different problems, so different functionals are used when calculating different problems. What problem fits what functional. For example, for a large organic conjugated system, it is reasonable to use PBE0 to optimize the ground state structure and CAM-B3LYP to calculate the electronic spectrum. Because PBE0 is recognized as suitable for optimizing the structure of the ground state of organic systems, this is also fully proved in the cross-measurement of J. Chem. Theory Comput., 12, 459 (2016). However, PBE0 is often not ideal for the calculation of electronic spectra of large conjugate systems (caused by low HF components), while CAM-B3LYP does not have this problem. It has been used by a large amount of literature to calculate the electronic spectra of large conjugate systems and has obtained good results. the result of. But for the ground state geometric structure optimization problem, CAM-B3LYP is not as good as PBE0, and it is rarely done in the literature. In addition, CAM-B3LYP is a range separation functional, and it takes more time than PBE0, so the ground state does not need to be optimized by CAM-B3LYP. (2) The time-consuming of the A functional is obviously lower than that of the B functional, and the accuracy is obviously worse than that of the B functional. Therefore, A functional is used to calculate those tasks that require low calculation level and a large amount of calculation, and B functional is used to more accurately calculate those problems that require high calculation level, that is, good steel is used on the blade. The most typical case is to use cheap functionals for optimization, and use significantly better and significantly more expensive functionals to calculate single point energy. For example, to calculate weak interactions, we can use B3LYP-D3 to optimize the structure, and use the dual hybrid functional PWPB95-D3, which is significantly higher in accuracy and time-consuming, to calculate single-point energy. (3) Related to research content and purpose. For example, if you want to compare and study the calculation accuracy of a series of functionals A, B, C, D for certain types of molecules such as polarizability, NMR, and dipole moment, then it is not necessary to use these four functionals to optimize the structure separately. With the calculation properties, you only need to use a functional that is reasonable for the current system to optimize, and then switch to different functionals to calculate separately when calculating the properties.

We consider a situation, B3LYP optimization, M06-2X is a single point, assuming that the current research is an organic system. In fact, there is a certain reason for the combination of these two functionals, because the time-consuming of M06-2X is higher than that of B3LYP (especially in vibration analysis), and the accuracy is higher overall. Moreover, M06-2X requires much higher integration grid points than B3LYP. If M06-2X is used for optimization and vibration analysis, if higher integration grid accuracy is not used (for example, the ultrafine grid points in Gaussian), false frequencies, The probability that geometric optimization is difficult to converge is quite large, and the use of higher integration grid accuracy will further increase the time-consuming of M06-2X. Therefore, such switching of functionals does have practical significance. But if you want to reduce the possibility of being made difficult by some reviewers, it is still recommended to unify the functionals, that is, use M06-2X for both optimization and single point calculations. After all, for organic systems, the geometric optimization of M06-2X is slightly better than B3LYP, especially when there are weak intramolecular interactions (of course, B3LYP can be corrected by adding DFT-D to make up for this), not to mention M06 After all, the calculation time of $-2X$ is still on the same level as B3LYP, and the gap is far less than that between ordinary functionals and double hybrid functionals. Of course, it is not to say that the author does not recommend using B3LYP optimization and M06-2X is a single point combination, but when doing so, you should clearly explain the reason for switching functionals in the article, otherwise it will easily make some readers and reviewers suspicious. As for the aforementioned optimization of PBE for transition metal systems, M06 is considered a single point. This approach is not advisable because it has no meaning for switching. The results of M06 for such systems may not be more reasonable than PBE, and it is time-consuming. It will improve. Therefore, if you want to use PBE, you must use PBE, and if you want to use M06, you must use M06. If you want to achieve better results, you should use double hybrid when you want to use a single point.

For Gaussian, although the density fitting method is supported to accelerate the calculation of pure functionals, it is very poorly done and no one generally uses them. Therefore, in Gaussian, the time-consuming of pure functionals and hybrid functionals is basically the same. However, in ORCA, Turbomole and other programs, because the density fitting of DFT is very good, pure functionals are more than an order of magnitude less time-consuming for large systems than hybrid functionals, so if you are a user of these programs, It is recommended to use pure functionals to fully save the calculation time of time-consuming steps. For ORCA users, for example, using BLYP-D3 (BJ) for optimization and vibration analysis, while using B3LYP-D3 (BJ) or M06-2X as a single point, this is completely acceptable and appropriate. But if you are a Gaussian user, it is not appropriate to switch functionals like this, and there is no justification (and in most cases, the accuracy of pure functionals is worse than hybrid functionals). Pure functionals have some significant weaknesses, such as the calculation of (hyper)polarizability will be significantly overestimated, the calculation of excitation energy will be significantly underestimated, and so on. For these tasks, you should never use pure functionals for the sake of cheapness.

Some people switch functionals when calculating different problems not because of the amount of calculation, nor the consideration of calculation accuracy, but for the purpose of collecting experimental data. They feel that the closer the calculation result is to the experiment, the better and easier to publish. In fact, this approach is not good. After all, theoretical calculations are not a foil for experiments, and there is no need to deliberately cater to experiments. Blindly let the results fit the experiment, what is the point of doing theoretical calculations? Every kind of functional has errors in the calculation of a certain problem in a certain type of system. As long as the selected functional is suitable in principle and all elements of the calculation process have been fully considered, certain errors should be accepted frankly, as long as the article The theoretical calculations can analyze and discuss the problems to be investigated clearly. As we all know, the higher the HF component of the functional, the higher the calculated excitation energy, so many people take advantage of this trend and choose a functional with the excitation energy closest to the experimental value to calculate the excitation energy, and then write it in the article To be proud, the calculation results in this paper are in good agreement with the experiments. This approach is really meaningless, and if you use a functional that is rarely used on the current problem for this purpose, then your intention to collect data is likely to be seen through by experienced reviewers. For example, if you calculate the UV–Vis spectrum of an organic system, you used B3LYP (20% HF content) when you optimized it before, but you may find that B3LYP slightly overestimates the excitation energy, so try to use a lower HF content (10%) TPSSh functional calculation, and found that the error of the result is a little smaller than that of B3LYP compared to the experiment, so I used TPSSh to calculate the spectrum in the article.

## 2.4  Basis Sets

### 2.4.1  Selection of Basis Sets

What basis set should I use? Before considering this issue, the following elements must be clarified and judged comprehensively. When asking what basis set the master uses, the more sufficient the following information is provided, the more accurate the master can give you (I am often asked what basis set to use, this is simply impossible to answer accurately)

(a) What kind of system, how many atoms, and what elements

(b) What is the problem

(c) How high precision is required and what theoretical method is used

(d) What computing resources can be used (especially the number of cores) and how long can be accepted for computing time

(e) What program to use.

This section first talks about how to choose the basis set, and then Sect. 2 gives specific selection suggestions for certain types of problems. The basis set selection recommendations given in this article are for the most commonly used Gaussian. Many of the basis sets mentioned in this article are not included in Gaussian, so you need to go to the BSE (https://www.basissetexchange.org) basis set database and other places to copy the definition.

**Basic Set Selection for General Problems**

Differences in the selection of time base sets for different methods. Note that different methods have completely different requirements for the basis set, which are reflected in the following two points: Dependence on basis function angular momentum and zeta number:Ordinary functional DFT and HF calculations (referring to these two cases when directly referred to as "DFT" below) do not require too many high angular momentum basis functions, while s and p angular momentum basis functions are significantly more meaningful for improving the results. Therefore, for B3LYP, the 3-zeta basis set 6-311G* is significantly improved compared to the 2-zeta basis set 6-31G*, and 6-311G (3df, 2pd) is more time-consuming than 6-311G (2d, p) There will be no improvement. Some people actually use the basis set of 6-31G(2df,p) to do DFT calculations, which is obviously extremely absurd. Even 3-zeta has not been reached. The areas that need to be improved are not improved, but increasing the amount of calculation has little effect on the DFT results. In terms of f polarization, there is a very lack of theoretical knowledge at first glance.

Then the post-HF, double-hybrid functional, and multi-reference methods (referring to these three cases when directly referring to the "post-HF method" below) have higher requirements on basis sets, and more high-angular momentum basis functions are required on the basis of DFT. In order to fully describe the related effects of electronic dynamics. Therefore, for high-level post-HF methods such as

CCSD(T), 6-311G**, which lacks sufficient high-angular momentum basis functions, is really a flower inserted in the cow dung. Use the advanced post-HF method at least from def2-TZVP, even if the low-level HF method Methods such as MP2 also have to start with at least def-TZVP.

The focus of the MCSCF method is not to describe electronic dynamic correlation but static correlation, and the requirements for the basis set are the same as those for DFT. The relationship between time consumption and the number of basis functions and GTF (Gaussian functions): Most of the time-consuming DFT calculation is used in the calculation of the two-electron integral, and the two-electron integral is formally proportional to the fourth power of the number of GTFs. Therefore, the number of GTFs in the system is the most direct influence on the time-consuming DFT calculation.

The time consuming of the post-HF calculation mainly depends on the number of basis functions (also related to the number of electrons in the system), because this directly determines the number of Slater determinants or configuration functions, and the cost of AO->MO integral transformation, while in the SCF iteration process The cost of dual electron integration is relatively minor.

In view of this, when the number of basis functions is the same, the basis set with a smaller number of GTFs is more popular for DFT, such as pople, def2, pcseg, these fragment shrinkage basis sets belong to this situation. The dunning-related consistency basis set (cc-pVnZ series) belongs to the generalized contraction. There are a large number of GTFs, and this basis set itself is optimized for post-HF electronic correlation calculations, so it is not cost-effective to use on DFT, but very suitable for use In the post HF calculation.

Recommendations for the selection of basic groups for general issues. For the DFT calculation of the non-double hybrid functional, according to the accuracy requirements, the general problem basis set selection recommendations are as follows. The symbol corresponds to the size relationship of the basis set in the same file. Dying struggling level: STO-3G (very small base) Deep water level: 3-21G (worst 2-zeta basis group) The lowest acceptable level: def2-SV(P) $\approx$ 6-31G* <def2-SVP $\approx$ 6-31G** $\approx$ pcseg-1 (2-zeta basis set) Good level: 6-311G** <def-TZVP (general 3-zeta basis set) Ideal level: def2-TZVP <def2-TZVPP $\approx$ pcseg-2 (high-end 3-zeta basis set) Invincible: def2-QZVP $\approx$ pcseg-3 (4-zeta basis set. It is wasteful to use for DFT).

For post-HF, double-hybrid functional, and multi-reference method calculations, according to the accuracy requirements, the general problem basis set selection recommendations are as follows. Minimum acceptable level: def-TZVP (general 3-zeta basis set) Good level: cc-pVTZ $\approx$ def2-TZVPP (high-end 3-zeta basis set) High-precision calculation: cc-pVQZ $\approx$ def2-QZVPP (4-zeta basis set) Invincible: cc-pV5Z (5-zeta basis set. It is wasteful and extremely expensive, usually through TZ or QZ CBS extrapolation to reach this level).

There is a consensus in academia that the pople basis set is not ideally constructed. For DFT calculations, the author strongly recommends using def/def2, pcseg series basis sets instead of pople basis sets under the same basis set size to obtain higher accuracy. For post-HF calculations, the pople basis set is an extremely poor choice, and it is strongly not recommended! The version of pople basis set larger than 6-311+G(2d,p) has no practical value at all, because there are other basis sets with much higher cost performance that can be used.

Single-point energy calculation, geometric optimization (including optimization to transition state), and vibration analysis are the most common types of tasks. Geometry optimization and vibration analysis are far less sensitive to basis set calculations than single-point energy calculations, and these tasks are much more time-consuming than single-point calculations, so geometric optimization absolutely does not need large basis sets, such as 6-31G* *There is no big problem with such a small basis set. Medium basis sets such as 6-311G** or def-TZVP are generally accurate enough. When the optimization is completed, the basis set will be increased by one. Two files are a wise approach, which belongs to the most basic common sense of quantum chemistry researchers. See http://sobereva.com/387 for details. IRC's requirements for basis set are the same as geometric optimization, and must be the same as when looking for transition states.

Don't forget to add a diffusion function if you need a diffusion function! What should be added in the aforementioned "Talk about Diffusion Function and "Month" Basis Set" has been clearly stated, and it will be mentioned later.

### Selection of Different Task Base Groups

For the calculation of excitation energy and electronic spectra (absorption, fluorescence, ECD spectra), the suitable basis set has been written in "The Calculation Method of Random Excitation State" (http://sobereva.com/265). What basis set is used has a certain relationship with the type of weak interaction. The nature of weak interaction is divided into two categories: diffusion effect: pi-pi accumulation, van der Waals attraction Electrostatic attraction is dominant, and a small amount of diffusion attraction participates: hydrogen bond, dihydrogen bond, halogen bond, pi-hydrogen bond, carbon bond, sulfur bond, phosphorus bond, etc. The greater the contribution of the diffusion effect, the more the diffusion function and the consideration of BSSE issues are needed.

The basis set recommendations for weak interaction energy calculation: Bottom limit: 6-31+G** or ma-def2-SVP for DFT; jun-cc-pVDZ for HF Return: use 6-311+G** for DFT; use may-cc-pVTZ for HF Good: use ma-def2-TZVP for DFT; use jun-cc-pVTZ for HF Ideal: aug-cc-pVTZ Extremely ideal: aug-cc-pVQZ Perfect: CBS. For weak interactions where diffusion dominates, if there is enough computing power, it is recommended to use counterpoise to deal with the BSSE problem when calculating the interaction energy. The time consumed will be more than twice that of the single point calculation of the complex.

The basis set recommendations for geometric optimization with weak interaction systems: diffusion-dominated weak interaction: may-cc-pVTZ is ideal enough, 6-311+G** is generally sufficient, 6-311G** or def-TZVP can also be accepted Electrostatically dominated weak interaction: def-TZVP is ideal enough, 6-31G** or def2-SVP are also acceptable.

It is strongly recommended to use pcSseg series basis set (JCTC,11,132) for NMR calculation, which is specially optimized for NMR calculation. pcSseg-1 is only a little bit larger than 6-31G**, but the NMR calculation results are similar to the much larger def2-TZVP and cc-pVTZ, which is quite ideal. The size of pcSseg-0 is similar to 3-21G. Although the accuracy is obviously not as good as pcSseg-1, it is worse than 6-31G** overall. If there is a margin in the calculation, it can be upgraded to pcSseg-2, but it is much larger than pcSseg-1 but the result is limited, which is not very cost-effective. If you have to use the common basis set to calculate NMR, you can use def2-TZVP, which can be reduced to def2-SVP if it doesn't work. IGLO-II is also good for NMR, but the recommendation is lower than the pcSseg series. It is not recommended to use the pople series basis set to calculate NMR, but must use at least 6-311G(2d,p).

If you need to calculate the spin-spin coupling constant J to study the splitting of NMR peaks, you must not use a general basis set to calculate this, you must use a dedicated basis set with a very tight (large exponential) basis function near the inner core. Row. It is strongly recommended to use the pcJ series, and the pcJ-1 results are very good. The IGLO-III basis set is also very suitable for calculating J.

Some people add diffusion to NMR calculations, which is stupid. Adding diffusion hardly improves the calculation accuracy of NMR.

When calculating NMR and J values, the atoms under investigation must not use pseudopotential basis sets, otherwise the shielding of the inner electrons from the external magnetic field cannot be shown at all. But using pseudopotential basis sets for adjacent atoms is completely okay.

The basis set of the barrier calculation can be as recommended in Sect. 1.2. If a good 3-zeta-level basis set has been used and the calculation volume is still rich, and you want to use it to further improve the accuracy, then it is recommended to add the lowest degree of diffusion (adding s, p angular momentum diffusion to heavy atoms), which will make The result is a quantitative improvement. It is worth mentioning that the improvement of the def2 series plus diffusion to the calculation results of the potential barrier is not as great as the pople basis set.

Do not add diffusion to the 2-zeta basis set when calculating the potential barrier. Using the energy of adding diffusion to increase the zeta number will significantly improve the accuracy.

Dipole moment, polarizability, and first hyperpolarizability are the first, second, and third derivatives of the external electric field of energy respectively. If you don't understand, please refer to "Analyze the output of Gaussian polarizability and first polarizability using Multiwfn" (http://sobereva.com/231). Calculating these three must have a diffusion function, otherwise the result is so bad that it is useless. And as the order of the derivative increases, the requirements for the diffusion function are getting higher and higher.

Calculate the dipole moment: If you use def2-TZVPD, there is no error at the basis set level. Using aug-cc-pVTZ is also good but much more expensive. If def2-TZVPD is not used, it will be reduced to aug-cc-pVDZ. If it is still not used, it will be reduced to def2-SVPD. If it is not used, then simply give up or use ORCA program to open RI. The Pople basis set calculates the dipole moment extremely badly, absolutely cannot use.

Calculating the polarization rate: According to the author's test (to be published), the most cost-effective basis set for this problem is sorted as ZPOL, jul-cc-pVDZ, aug-cc-pVDZ, POL, aug-cc-pVTZ(-f,-d), LPol-ds. Among them, ZPOL, POL, and LPol-ds are all proposed by Sadlej, see "Definitions of Gaussian Formats of Various Sadlej Basis Sets" (http://sobereva.com/345). Among them, aug-cc-pVTZ(-f,-d) is obtained by cutting off the f polarization of heavy atoms and the d polarization of light atoms of aug-cc-pVTZ. In addition, although the basis set with D suffix of the def2 series basis set such as def2-SVPD is also optimized for the polarizability calculation, it is not as cost-effective as the above basis set in my test, but it can be used if you want. If pseudopotentials are used, it is recommended to use the LFK pseudopotential basis set, which is modified on the basis of the SBKJC pseudopotential basis set, so that the calculation accuracy of the polarizability is similar to the all-electronic POL basis set.

For calculation of molecular vibration frequency, infrared spectroscopy, VCD spectra, for large systems, 6-31G* is actually sufficient for this type of task base group, and def-TZVP is sufficient at most. It cannot be increased any more. A waste of time. Generally speaking, these tasks are calculated based on resonance approximation, and frequency correction factors are generally used to correct the systematic errors of resonance approximation and theoretical methods. See "Talking about resonance frequency correction factors" (http://sobereva.com/221). After the frequency correction factor is corrected, there is no difference in the results with small basis set, medium basis set, and large basis set. Calculating this type of task with a large base group is done by a rookie.

Although Raman and ROA spectra are also vibration spectra, the calculation of them requires a diffusion function to get better results. Because they all involve the derivative of the polarizability with respect to the vibration coordinate when calculating, and the polarizability needs to have a diffusion function to calculate it accurately. J. Chem. Theory Comput., 7, 3323 (2011) proposes a two-step method for calculating Raman and ROA. No diffusion function is used in optimization and vibration analysis (as low as 6-31G* as high as def-TZVP), but When calculating the polarizability derivative, consider the diffusion function (recommendation >= aug-cc-pVDZ), which is cheaper than using the diffusion function in the whole process and has no loss in accuracy. Gaussian supports this two-step method. (Because ROA and Raman are vibration spectra, frequency correction factors should also be considered).

All-electronic relativity calculations need to use DKH, ZORA, X2C and other Hamiltonian considering relativistic effects. At this time, it is necessary to use a basis set specifically for relativity calculations to optimize the shrinkage method, such as cc-pVnZ-DK series, SARC, etc., or simply Use a completely unshrinkable basis set such as UGBS. What can be used at the beginning of this article briefly mentioned "Some Notes on Wave Function Analysis under Pseudopotentials" (http://sobereva.com/156). The ordinary basis sets mentioned in Sect. 1.2 must not be used for all-electronic relativity calculations.

Some beginners don't know anything about the calculation of all-electron relativity, and they used pseudopotential basis sets and used DKH2 relativity Hamiltonian, at this time the result is completely wrong! When using the theory of relativity, Hamilton must be an all-electronic basis set.

However, if you do not consider the scalar part of the relativistic effect and only consider the spin-orbit coupling part, you can use the usual all-electronic basis sets and pseudopotentials.

For wave function analysis based on real space functions, such as discussing electron density, ELF, electrostatic potential, AIM, RDG analysis, etc., the basis set 6-31G* is generally sufficient to get qualitatively correct results, and 6-311G is used ** or def-TZVP is more than enough. If it is an anion system, adding layers of s and p angular momentum diffusion functions is also sufficient. It is useless to increase the basis set, because these real space functions increase quickly with the basis set. Will converge. The NBO analysis has the same requirements for the basis set. Adding the diffusion function has almost no effect on the distribution of the real space function, so adding it will only waste calculation time, but it will not make the result worse.

For CDA analysis, Mayer key level, Wiberg key level, Mulliken analysis, SCPA analysis, as well as Multiwfn based on Mulliken/SCPA to draw PDOS, OPDOS graphs and other direct basis function analysis, the basis set will not be improved if it is used too large. As a result, the result under 6-311G** is by no means better than the result under 6-31G**. If you continue to raise the base set to 4-zeta, the result will only get worse. If the basis set has a diffusion function, then the result is useless at all! It doesn't work at all! It doesn't work at all! (It is very important, so I say it three times.) It is absolutely impossible to use basis sets with diffusion functions for these analyses, because the diffusion functions have no chemical meaning at all, so the results of these methods of dividing the atomic space directly based on the basis functions will be messed up.

To calculate the solvation energy using the implicit solvent model, you must use the special parameterized level of the solvent model. For example, the best level for SMD parameterization is M05-2X/6-31G*, so it is recommended to be at the level of M05-2X/6-31G* Combine SMD to calculate the free energy of dissolution. If the basis set is used larger, such as aug-cc-pVTZ, the result will only get worse.

For programs based on Gaussian functions such as Gaussian and Crystal, the 6-31G** basis set commonly used to calculate isolated systems is also suitable for organic periodic system calculations. But many basis sets dedicated to molecular calculations should not be used directly in periodic calculations, such as 6-311G* to calculate Si. Because these basis sets often have basis functions with small exponents and certain diffusion characteristics, which are useful for describing the tail characteristics of atomic orbitals, but in periodic systems, this is not required to be described, and it may cause numerical problems. Therefore, these The exponent of the basis function is artificially adjusted or removed. Some basis functions with redundant basis sets should also be appropriately removed. The basis set with its own diffusion function must not be used for periodic calculation, otherwise it will be very troublesome.

The basis set of Gaussian functions suitable for periodic calculation can be obtained from the Crystal website, but the format needs to be changed: http://www. crystal.unito.it/Basis_Sets/Ptable.html. The pob-TZVP is a modified version suitable for solid calculation based on def2-TZVP, which can get better results than other 3-zeta basis sets.

Fermi contact is the main source of the hyperfine coupling constant. It is necessary to calculate the spin density at the nucleus. It is recommended to use the EPR-II and EPR-III basis sets optimized for DFT calculations for hyperfine coupling (built in Gaussian, only for H, B, C, N, O, F are defined), or IGLO-III, pc-J series basis set calculations suitable for calculating the nuclear spin-nuclear spin coupling constant. This type of basis set has a relatively tight s-function, which can better describe the spin density at the nucleus.

A basis set optimized for explicit correlation calculations should be used, the most well-known is the cc-pVnZ-F12 series.

It is generally recommended to fit the basis set with a special matching density, such as the def2 series. If there is no matching density fitting basis set, such as the pople series, when using the density fitting calculation, you have to let the program automatically generate the density fitting basis set, which will take more time than the standard configuration, so it speeds up The effect is not so good. It is also possible to borrow the density of the basis set of similar level or higher to fit the basis set, but it is not perfect after all.


**Notice in Basis Sets**

**When to use pseudopotentials?** The main uses of pseudopotentials are: (1) describe the inner electrons that are not chemically interested in an equivalent potential field, so there is no need to explicitly express the inner electrons, which greatly saves the amount of calculation (2) can wait Effectively reflect the scalar relativity effect. The relativity effect of the fourth cycle begins to appear, but it is not a big problem if it is not considered; and the relativistic effect of the fifth cycle and later cannot be ignored, and the results are not considered or even qualitatively wrong.

Pseudopotential and pseudopotential basis sets are used together. Generally speaking, it is recommended to use the pseudopotential basis set from the fourth cycle, which not only saves time, but also can be equivalently included in the relativistic effect improvement results (provided that the relativistic pseudopotential RECP is used, that is, the original fitting pseudopotential RECP is used. The effect of relativity has been taken into consideration when the situation is). If you are interested, you can read the "Functional Forms of Pseudopotentials and the Ways of Definition in Quantum Chemistry Programs" (http://sobereva.com/188) and the aforementioned "Some Notes on Wave Function Analysis under Pseudopotentials", Pseudopotentials The selection of basis sets is discussed in detail in the aforementioned "Talk about the selection of pseudopotential basis sets".

**What elements does the basis set contain?** Only a few basis groups, such as UGBS and WTBS, cover almost the entire periodic table, and most basis groups only define a part of the periodic table. What elements are defined by the basis set can be viewed on the BSE website by clicking on the corresponding basis set.

**What does CBS mean?** CBS=complete basis set (complete basis set). Using CBS is equivalent to using an infinite basis set. At this time, the result error caused by the incomplete basis set is 0. CCSD(T)/CBS is often written in the literature. Of course, the basis set used in the actual calculation cannot be infinite. The result under CBS is achieved by extrapolating the result under the finite basis set. For details, see the aforementioned "Talk" "On the basis set extrapolation of energy."

**How to obtain the basis set?** Since the basis functions used in most quantum chemistry programs are Gaussian functions, and the mathematical form is consistent, the basis set of Gaussian functions can be used in any quantization program based on Gaussian functions. Basically all quantum chemistry programs have their own basis set libraries, including mainstream basis sets, which can be used directly.

However, the built-in basis sets are often not comprehensive enough, and the basis set definitions often need to be obtained from other places to use in the program. It can be borrowed from other program base set libraries, or handwritten based on the data in the original document of the base set, or obtained from the website of some base set developers. The most common way to obtain an external basis set is through the BSE basis set database, which is the most complete public basis set database (very few basis sets may have errors in the data of individual elements). For details, see the aforementioned "List of Online Basis Sets and Pseudopotential Databases".

**About mixed basis sets** Mixed basis set means that some atoms use a certain basis set, and some atoms use another basis set. Appropriate use of mixed basis sets can greatly reduce the calculation time without much loss of accuracy. It is recommended that you make full use of this approach. For example, to calculate the chemical reaction process of a large system, the reaction part only involves one of the regions, then only the large basis set (such as def2-TZVP) can be used for the atoms in this region and the middle/small basis set (such as 6-31G* or even 3-21G). For another example, if there is a large amount of negative charge locally in the system, it is not necessary to add a diffusion function to all the atoms in the entire system. It is enough to add a diffusion function to only that part of the atoms. In the calculation of complexes, pseudopotential basis sets are usually used for transition metal atoms and all-electron basis sets are used for light atoms. This actually belongs to the use of mixed basis sets.

For the method of using mixed basis sets in Gaussian, please refer to the aforementioned "Detailed explanation of the input of mixed basis sets, custom basis sets and pseudopotential basis sets in Gaussian".

**BSSE issues** When the Gaussian function basis set is actually used, it is basically centered on the atom. The basis function of a certain atomic band will invade the space of the nearby atoms, which leads to BSSE (The basis set overlap error), when calculating the interaction energy, the effect intensity will be overestimated. There are counterpoise, gcp and other methods to solve the BSSE problem. It is especially emphasized that the counterpoise is only suitable for energy calculation. It is strongly not recommended to use it during optimization and vibration analysis. Otherwise, there will be no analytical derivative, which is too slow for large systems to vomit blood, and Generally, it does not bring any detectable improvement.

**Whether to add polarization function or diffusion function to hydrogen?** Calculations that are closely related to hydrogen, such as the calculation of hydrogen bonds, hydrogen transfer reactions, protonation energy, hydrogen NMR spectrum, etc., are necessary to add polarization functions to hydrogen, and other conditions will not cause hydrogen to add polarization. What an error.

Adding a diffusion function to hydrogen is usually of low significance. It is specifically explained in "Talking about the diffusion function and the "month" basis set", and will not be repeated here.

**Which tasks must use the same basis set?** Many beginners have heard some misunderstandings on the Internet and are confused about the concept. They often ask whether the basis set used for calculation must be consistent with the basis set used for optimization. In fact, only the level used for IRC and vibration analysis (obviously including the basis set used) must be consistent with the geometric optimization (including optimization of transition states). In addition, any common tasks are absolutely absolute.

**About the basis set of the Pople series plus the polarization function** One feature of the Pople series basis set is that it is highly customizable. You can control how much and what angular momentum to add to the polarization function behind 6-31G and 6-311G. For heavy atoms, it can be d, df, 2d, 2df, 3d, 3df, 3d2f, for light atoms can be p, pd, 2p, 2pd, 3p, 3pd, 3p2d. Many people add randomly and do not pay attention to the balance of angular momentum. For example, it is appropriate to gradually increase in this order and try to improve the result: (d,p) (2d,p) (2df,2p) (2df,2pd) (3d2f,2pd) (3d2f,3p2d). But if you use (3d, 3p2d), it is very inappropriate. There are so many polarizations for heavy atoms d. Why is there not a single f? Moreover, with so much polarization of hydrogen, the polarization of heavy atoms appears to be less. In addition, for the 2-zeta basis set of 6-31G, it is enough to give (d, p) or at most (2d, p). If you use 6-31G (2df, 2pd), it will be stupid. Using so many polarization functions to improve the results is far better than upgrading the basis set to 3-zeta.

In addition, the pople basis set is mainly suitable for crude or moderate HF/DFT calculations, even the post HF of general accuracy is not suitable for the pople basis set (suitable for dunning related consistency basis sets). 6-31G*, 6-31G**, 6-311G** can be used for DFT, but if you want to improve the accuracy, then it is strongly recommended not to use pople basis set, and add more polarization functions to it by yourself It is far inferior to directly using the base set (def2, pc series) specially matched by others. For example, def2-TZVP and 6-311G (2df, 2p) are time-consuming, and the former is obviously better than the latter.

By the way, 6-31G and 6-311G are useless without polarization function. Compared with 6-31G, 6-31G* and 6-311G* have an accuracy of one sky and one underground compared to 6-311G. Although the polarization is more time-consuming, it is super cost-effective. I often see that beginners dare to use the bare 6-31G and 6-311G directly.

**The problem of misusing the diffusion function** When should the diffusion function be used? I specifically talked about it in the aforementioned "Talking about the diffusion function and the "month" basis set". This article also mentioned many times when to add and when not to add the diffusion function. The old man has been answering computational chemistry questions on the Internet for a long time, and there are too many cases where others use the diffusion function indiscriminately. Again, if you have read the above and "Talking about the Diffusion Function and the "Month Basis Set" carefully, and you are not sure whether you should add a diffusion function, then the answer is: don't add a diffusion function. The situation where the diffusion function is added when it should not be added is at least 10 times more than the situation where the diffusion function should be added but not added.

Adding diffusion when the diffusion function shouldn't be added almost does not improve the results at all, and it also causes many problems, especially the sudden increase in calculation volume and difficulty in convergence. You know, the same is to add a layer of sp basis function. If you add a diffusion function, it will take much more time than increasing the number of zeta! That is, for heavy atoms, compared with 6-31G*, the time-consuming increase to 6-31+G* is significantly higher than that of 6-311G* (the larger the system, the more obvious this phenomenon is). Because the index of the diffusion function is small and the extension range is wide, it is not easy for the quantization program to rely on the integral screening strategy to ignore the integral that contributes little to the result to save the amount of calculation, and the diffusion function itself also easily leads to the need for more turns for convergence. And for the case where a diffusion function is not necessary, 6-311G* is better than 6-31+G*. Even for problems that are more useful for diffusion functions, such as the optimization of diffusion-dominated weak interaction systems, considering the amount of calculation and task type, I recommend 6-311G* instead of 6-31+G*, because the former itself is at this time The result is not worse than the latter. Although it does not have a diffusion function, the outermost layer of 3-zeta itself shows a little bit of diffusion, which is basically enough. Moreover, the accuracy of 3-zeta's description of valence layer electrons is also better than that of 2-zeta.

## 2.4.2  Application of Mixed Basis Set, Custom Basis Set and Pseudopotential Basis Set in Gaussian

### Input of Mixed Basis Set

Mixed basis set means that different atoms use different basis sets. Writing the gen keyword means reading the basis set definition from the back of the coordinate part. Use IOp(3/24=1) to display the basis set actually used by each atom to check whether it is set correctly; you can also use the GFPrint keyword to output similar content, but it is not so clear; you can also use the GFInput keyword to output the actual The basis set used, these information can be directly used as the input information of the basis set.

The following example is CH4. For example, if you want C to be 6-311G* and H to be 6-31G**, the input file is

```
   m062x/gen
   [Blank line]
   test
[Blank line]
0 1
C -0.00000000 0.00000000 0.00000000
H -0.00000000 0.00000000 1.09000000
H -0.00000000 -1.02766186 -0.36333333
H -0.88998127 0.51383093 -0.36333333
H 0.88998127 0.51383093 -0.36333333
[Blank line]
C 0
6-311G*
****
H 0
6-31G**
****
```

It must be ensured that each atom defines a basis set. In addition, it is not possible to define more basis sets. For example, change C 0 to C F 0 to indicate that F also uses 6-311G* basis sets. Since there is no F in the system, an error will be reported. But you can write-in front of the element symbol, which means that if there is this element in the molecule, the basis set definition will take effect. If there is no such element, no error will be reported. For example, you can change the above C 0 to C -F 0, which is equivalent to adding this item at the end.

```
   -F 0
6-311G*
****
Will not cause errors.
```

If you only want H2, H3, H4 to be 6-31G**, and H5 to use 6-311G*, you cannot write:

```
    C H 0
6-311G*
****
2  3  4  0
6-31G**
****
```

The consequence of this is that H2, H3, and H4 are assigned the basis functions defined by 6-311G* and 6-31G** at the same time, which is wrong. Because the basis set is defined repeatedly for the same atom, the basis function is an additive relationship, not a covering relationship. In order to achieve the goal correctly, it should be written like this:

```
    C  5  0
6-311G*
****
2  3  4  0
6-31G**
****
```

**The Form of Custom Basis Set**

The custom basis set mentioned here refers to the input and modification of the specific definition of the basis set by oneself, instead of directly using the ready-made basis set built in Gaussian. This requires a basic understanding of the definition of the basis set.

The definition of basis set can be obtained from the BSE basis set database, that is, go to https://www.basissetexchange.org, click on the element symbol, click on the basis set you want to use on the left, then select Gaussian94 as Format, and then click the Get Basis Set button, Just copy the data inside. Gaussian also has a base set library, which can be seen directly for the Linux version, which is the basis subdirectory of the Gaussian directory. For example, when you open 6311.gbs with a text editor, you will see the definition of various elements in 6-311G, and 6311s.gbs records the definition of various elements in the polarization function of 6-311G. Compared with the basis set definition copied from BSE, you will find that it is consistent.

Here we take the 6-31+G* basis set of carbon atoms as an example to illustrate the definition format of basis set in Gaussian.

```
    C  0
S  6  1.00
3047.5249000  0.0018347
457.3695100  0.0140373
103.9486900  0.0688426
29.2101550  0.2321844
```

```
9.2866630 0.4679413
3.1639270 0.3623120
SP 3 1.00
7.8682724 -0.1193324 0.0689991
1.8812885 -0.1608542 0.3164240
0.5442493 1.1434564 0.7443083
SP 1 1.00
0.1687144 1.0000000 1.0000000
SP 1 1.00
0.0438000 1.0000000 1.0000000
D 1 1.00
0.8000000 1.0000000
****
```

When using a custom basis set, also write the gen keyword, and then write a blank line at the end of the molecular coordinates and then write the basis set definition in the above form. Here we copy the definition of STO-3G from BSE to methane

```
# m062x/gen
[Blank line]
test
[Blank line]
0 1
C -0.00000000 0.00000000 0.00000000
H -0.00000000 0.00000000 1.09000000
H -0.00000000 -1.02766186 -0.36333333
H -0.88998127 0.51383093 -0.36333333
H 0.88998127 0.51383093 -0.36333333
[Blank line]
H 0
S 3 1.00
3.42525091 0.15432897
0.62391373 0.53532814
0.16885540 0.44463454
****
C 0
S 3 1.00
71.6168370 0.15432897
13.0450960 0.53532814
3.5305122 0.44463454
SP 3 1.00
2.9412494 -0.09996723 0.15591627
0.6834831 0.39951283 0.60768372
0.2222899 0.70011547 0.39195739
****
```

## Combination of Custom Basis Set and Mixed Basis Set

In fact, Gaussian will first expand each element symbol into an atomic number, and expand each basis group name into specific information about the basis group. The basis set definition part of the example in the first section can also be equivalently written as follows, where the specific definition of the 6-311G* basis set of C is directly written, and H is the basis set name directly.

```
 C 0
S 6 1.00
4563.2400000 0.00196665
682.0240000 0.0152306
154.9730000 0.0761269
44.4553000 0.2608010
13.0290000 0.6164620
1.8277300 0.2210060
SP 3 1.00
20.9642000 0.1146600 0.0402487
4.8033100 0.9199990 0.2375940
1.4593300 -0.00303068 0.8158540
SP 1 1.00
0.4834560 1.0000000 1.0000000
SP 1 1.00
0.1455850 1.0000000 1.0000000
D 1 1.00
0.6260000 1.0000000
****
H 0
6-31G**
****
```

1G* basis set of C is directly written, and H is the basis set name directly.

## Add Additional Functions to Existing Basis Sets

The situation where additional functions are needed is mainly used to add diffusion and polarization functions.

For the example of CH4 in the first section, if you add a diffusion function to C that originally used 6-311G*, of course, the easiest way is to rewrite it as 6-311+G*, but you can also customize the basis set Way to add a diffusion function to it. We can find the part corresponding to the diffusion function in the 6-311+G* basis set of C from the BSE, or from the plus.gbs file in the basis directory (the definition file of the diffusion function of the heavy atom of the Pople basis set) Copy that part of the information of C and add it to the end of the 6-311G* basis set definition of C,

```
 C  0
6-311G*
SP  1  1.00
0.4380000000D-01  0.1000000000D+01  0.1000000000D+01
****
H  0
6-31G**
****


It can also be written like this, which is equivalent
to the above, because the basis set of C is defined
twice, which is equivalent to adding together
C  0
6-311G*
****
H  0
6-31G**
****
C  0
SP  1  1.00
0.4380000000D-01  0.1000000000D+01  0.1000000000D+01
****
```

Similarly, we can add polarization functions to specific atoms.

Many documents give the index of the polarization/diffusion function added to a certain basis set for a certain element. Since we already know what the shell type is, and that its shrinkage must be 1, and the shrinkage coefficient and scale factor are both 1.0, we can directly add these polarization/diffusion to the standard format of the custom basis set In the current basis set.

PS: When defining the basis set, you will often see the suffix Dxxx like above. This is the representation of double-precision floating-point numbers used by Fortran programs such as Gaussian. D is equivalent to E in scientific notation. For decimals, it is strongly recommended to always write in double-precision notation, because Gaussian uses double-precision floating point. If you write 0.0438 directly in the basis set definition, the Gaussian will be read as a single-precision floating-point number and read into a double-precision variable. Because the effective digits of single-precision floating-point are relatively limited, it may become For example, 0.04380003215, some extra "noise" at the end affects the calculation accuracy.

In addition, there is an extrabasis keyword in Gaussian. The difference between it and gen is that gen completely redefines the basis set of all atoms, while the extrabasis keyword is only used to add some additional basis functions to the current basis set.

You only need to define additional Just add those basis functions. Assuming that the current 6-31G* is used, we want to add a diffusion function to the carbon, we can write the keyword part as # M062X/6-31G* extrabasis, and then write a blank line at the end of the molecular coordinates

```
C 0
SP 1 1.00
0.4380000000D-01 0.1000000000D+01 0.1000000000D+01
****
```

It can be seen that the obvious difference between extrabasis and gen when used to add additional functions is that the former does not need to rewrite the original basis set after the molecular coordinates, but the disadvantage is that the original basis set cannot be a mixed basis set, because it cannot be combined with gen Mixed together, it is impossible to define the original mixed basis set at the same time.

**Tips: Use File References to Set the Basis Set**

As mentioned earlier, the .gbs file in the basic folder of the Linux version can find the definition of various elements of various built-in basis sets of Gaussian, and the corresponding basis set can be roughly guessed through the file name. For example, 631.gbs corresponds to 6-31G*, and the content is:

```
-H
S 3 1.00
0.1873113696D+02 0.3349460434D-01
0.2825394365D+01 0.2347269535D+00
0.6401216923D+00 0.8137573262D+00
S 1 1.00
0.1612777588D+00 0.1000000000D+01
****
-He
S 3 1.00
0.3842163400D+02 0.2376600000D-01
...(slightly)
```

In fact, the information of these files can not only be consulted and copied, but also can be directly entered into the input file through the file reference function of Gaussian. E.g:

```
{Molecular coordinate information}
{Blank line}
@/sob/g03/basis/sto3g.gbs
```

where @ represents a reference to an external file. In Gaussian's Link0 module, this entry will be expanded into the actual file content, just like include in programming. Of course, you can copy the contents of the .gbs file directly to the input file. There are many elements defined in .gbs, and they will certainly not appear in the current molecule at the same time. Since each element in this file has a negative sign before the symbol, those elements that are defined more will not cause an error, but will be ignored.

Using Gaussian's file citation method will bring great convenience to actual research. For example, if you want to study a large number of molecules, you must use a custom basis set, so you don't need to define the basis set in every file. Once again, instead of directly referencing a file that records the definition of the basis set. In the article "A basis set definition file in Gaussian format (including all elements supported by def2) that add a diffusion function to def2 in a ma- manner" (http://sobereva.com/509), it is also demonstrated to reference external files to facilitate Use ma-def2 series basis sets.

### Use Pseudopotential Basis Sets

The basis set of pseudopotentials needs to be used together with the corresponding pseudopotentials. If you are not familiar with pseudopotentials or want to understand the inner details, please refer to "Functional Forms of Pseudopotentials and the Ways of Definition in Quantum Chemistry Programs" (http://sobereva.com/188). To use the pseudopotential basis set, you need to use the genecp keyword (equivalent to gen pseudo=read), which means first read the pseudopotential basis set definition from the molecular coordinates, and then read the pseudopotential definition. For example $Cu(CO)+$:

```
# B3LYP/genecp
[Blank line]
Cu(CO)+
[Blank line]
1 1
Cu
C 1 B1
O 2 B2 1 180.
[Blank line]
B1 1.94000000
B2 1.11540000
[Blank line]
Cu 0
Lanl2DZ
****
C O 0
6-31G*
```

```
****
[Blank line]
Cu 0
Lanl2
[Blank line]
[Blank line]
```

The usage of Stuttgart pseudopotential is relatively special, so I will mention it here. The stuttgart pseudopotential specification is written in the form nXY, which means n = the number of nuclear electrons represented by the pseudopotential X = S/M: The reference system when fitting pseudopotentials. S is the energy of only a single valence electron in the model system, and M is the energy of all valence electrons in the actual atom Y = HF/WB/DF: The calculation level of the data used to fit the pseudopotential. HF=Hartree–Fock non-relativity (that is, this pseudopotential does not consider the effect of relativity), WB=Wood-Boring quasi-relativity, DF=Dirac–Fock total relativity Gaussian comes with these combinations: SDF2, MWB2, SDF10, MWB10, MDF10, MWB28, MWB46 60, MWB78. Not all of these elements can be used. For example, P can only use MWB10 and SDF10, La can only use MWB28/46/47 and MHF46/47 (MWB28 is a small nuclear pseudopotential for La, and the others are large nuclear pseudopotentials.), see the list of pseudo keywords in the manual for details. The writing of these Stuttgart pseudopotentials is often not easy to remember. In Gaussian, people are used to using SDD keywords. Write SDD to indicate that D95V or 6-31G* all-electronic basis sets are used for elements with sequence number <=Ar, and Stuttgart pseudopotential and corresponding pseudopotential basis sets are used for heavier elements. Which one of the nXY notation is used specifically Species, see the table of pseudo keywords. In addition, there is another keyword SDDAll, which is different from SDD in that it uses Stuttgart pseudopotential and corresponding pseudopotential basis set for all elements with sequence number > 2, instead of D95V for light elements. The Stuttgart series of pseudopotentials are constantly being developed and improved, and the Stuttgart pseudopotentials built into Gaussian (including those on the BSE) are neither complete nor new. The latest version can be downloaded from the website of the developer of the Stuttgart series of pseudopotentials: http://www.tc.uni-koeln. de/PP/clickpse.en.html.

An example of using Stuttgart's pseudopotential:

```
Cu 0
SDD
****
C O 0
6-31G*
****
[Blank line]
Cu 0
SDD
```

According to the table in the pseudo keyword section of the manual, the two SDDs can also be written as MDF10, which are equivalent.

Different pseudopotentials can be used for different elements, and you can write multiple in the pseudopotential definition part. For example, use MWB46 for lanthanum and MWB60 for lutetium, which should be written as

```
C H O N 0
6-31G**
****
La
MWB46
****
Lu
MWB60
****
[Blank line]
La 0
MWB46
Lu 0
MWB60
[Blank line]
[Blank line]
```

Using the method described above, additional functions can also be added to the pseudopotential basis set. A very common situation is to add a polarization function to the Lanl2DZ pseudopotential basis set. For Cu, adding polarization refers to adding the f function. Its exponent is scattered in some literatures, and it can also be borrowed from the Lanl2TZ(f) basis set of Cu on the BSE. The value is 3.525. After adding f polarization, the content after the molecular coordinates becomes

```
Cu 0
Lanl2DZ
F 1 1.0
3.525D0 1.0D0
****
C O 0
6-31G*
****
[Blank line]
Cu 0
Lanl2
[Blank line]
[Blank line]
```

Sometimes we need to use the pseudopotential basis set that Gaussian does not have built-in. For example, Lanl08, Lanl2TZ, cc-pVnZ-PP series, etc. At this time, we need to copy the definition from the BSE and paste it to the pseudopotential basis set and the place of the pseudopotential definition. For example, using cc-pVDZ-PP for Cu in the above example, go to BSE to find the corresponding entry. The first half of the information displayed on the page is the definition of the pseudopotential basis set, and the second half is the definition of the pseudopotential. After using this example, the molecular coordinates follow Part of it becomes

```
Cu 0
S 7 1.00
560.0880000 0.0006370
56.6486000 -0.0097350
35.4258000 0.0657930
11.0546000 -0.4150350
2.3068200 0.7466110
0.9514290 0.4621730
0.1451840 0.0159830
...(slightly)
D 1 1.00
0.2836420 1.0000000 ,
F 1 1.00
2.7482000 1.0000000
****
C O 0
6-31G*
****
[Blank line]
CU 0
CU-ECP 4 10
g-ul potential
1
2 1.0000000 0.0000000
...(slightly)
f-ul potential
2
2 6.1901020 -0.2272640
2 8.1187800 -0.4687730
[Blank line]
[Blank line]
```

The basis set of the def2 series is also quite special, so I will talk about it here. This series of basis sets are all-electron basis sets for the first four cycles (H Kr), and pseudopotential basis sets for the fifth and sixth cycles (excluding actinium), with the small nuclear Stuttgart pseudopotential. When using the def2 series in Gaussian and involving elements of the fifth period and later, you do not need to use genecp to specifically define pseudopotentials for the elements of the fifth period and later. For example, to optimize OsO4, just write B3LYP/def2SVP opt. At this time, Gaussian automatically knows to use the def2SVP all-electronic basis set for O and the def2SVP pseudopotential basis set for Os and add the corresponding pseudopotential. So it is very convenient to use the def2 series.

### 2.4.3 Diffuse Functions

**When Is a Diffusion Function Needed?**

The diffusion function in quantum chemistry calculation refers to a basis function with a small exponent and a wide spatial distribution range. The necessity of adding a diffusion function, the author summarizes the following based on a large number of theoretical calculation articles and practical experience:

- The result without diffusion function must be useless:

  Calculate dipole moment, multipole moment
  Calculate polarizability and hyperpolarizability
  Calculate Rydberg Excited State
  Calculate the energy and electron affinity of the anion system

- Adding diffusion function is very important, it is strongly recommended to add diffusion:

  Calculate weak interaction energy.
  Calculate the strength of Raman and ROA.

- Adding diffusion function is beneficial to improve the results, and the calculation conditions allow it to add diffusion:

  Calculate the reaction barrier.
  Optimize the structure of anion system or weak interaction system and perform vibration analysis on it.

If the system has only part of the atoms with very significant negative charges or its electrons are easily polarized, or the weak interaction involves only a part of the system, and limited computing power cannot add diffusion to all atoms, then at least these regions The atom plus the diffusion function.

When the diffusion function is indispensable, it is better to reduce the zeta number in order to be able to add the diffusion function. For example, when calculating the dipole moment and polarizability, it is better to reduce the diffusion function from 3-zeta to 2-zeta. But when the diffusion function does not play a key role but it is beneficial to add it, then at least use the basis set to a grade no lower than 3-zeta and then consider adding the diffusion, this is the most cost-effective, because at this time increase the basis set The benefit brought by the number of valence layer splits is obviously much greater than adding the diffusion function.

For a neutral system where no significant weak interaction is involved, and no local significant negative charge is involved, calculate the characteristics not mentioned above, such as atomization energy, valence electron excitation, bond energy, ionization energy, geometric optimization, NMR, etc., do not add diffusion, otherwise it will not improve the accuracy of the results, which will increase the calculation time by a huge amount and cause many problems. Remember that using the diffusion function indiscriminately will only bring itself shame!

PS: I see from time to time that some people add a diffusion function when calculating cations (such as energy and geometric optimization). It is simply incomprehensible. The cation does not have weakly bound electrons like anions. What diffusion function should be added!? The reviewer knew at a glance that the author lacked common sense of computational chemistry, and his favorability dropped sharply!

Since it is very common to misuse the diffusion function for hydrogen, I will specifically talk about it. Adding a diffusion function to hydrogen is usually meaningless, because hydrogen itself has only one electron, and its electronegativity is small, and a lot of electrons are generally absorbed in a molecular environment, so adding a diffusion function to it obviously has no effect. For organic systems with a large number of hydrogens, the amount of calculations increases a lot. Therefore, when it is necessary to add a diffusion function to the system, it is not recommended to use such as 6-311++G**, but 6-311+G** is recommended. Moreover, even if you study the hydrogen bond energy, you don't need to add a diffusion function to hydrogen. Tests show that this has little effect on the results. But for the calculation of (super) polarizability, hydrogen with obvious negative charge (such as LiH), etc., adding a diffusion function to hydrogen is indeed of great significance. At this time, it is necessary to add diffusion to hydrogen.

## Problems Caused by Diffusion Function

The addition of the diffusion function will increase the completeness of the basis function. In principle, it seems that it should always be beneficial and harmless to the result, but in actual calculations it will cause the following problems: The amount of calculations skyrocketed. This is well-known, especially the cc-pVnZ series, which often doesn't move even after adding aug-. The fourth section will discuss how to solve this difficulty. SCF is often much harder to converge after adding the diffusion function than without adding it. The chemical significance of the diffusion function is very poor, and the lack of correspondence with the atomic orbital will cause serious adverse effects on the wave function analysis method in Hilbert space. For example, the Mulliken charge will become extremely bad, and the Mayer key level will be quite unreliable. The reason is not difficult to understand. For example, a large number of diffusion functions of A atom extend into the space of B atom, so a part of the electron distribution near B will be described by these diffusion functions, then Mulliken population analysis will take a lot of what should belong to B The electron is assigned to the A atom, causing the charge of A to be too negative and B to be too positive. The chemical meaning of virtual orbitals (that is, non-occupied orbitals) becomes more ambiguous. Especially in the Hartree–Fock calculation under the diffusion function, the spatial distribution range of the virtual orbit tends to become extremely wide, which makes the frontline orbit theoretical analysis no longer applicable.

The symmetry of the system structure and wave function may decrease due to the accuracy of the numerical calculation. For example, if you optimize a system with symmetry, the initial structure program can determine the actual point group, but after optimization, the program can sometimes only determine the lower-order point group. If you are currently using a diffusion function, 80% is caused by the diffusion function. Removing the diffusion function can often maintain symmetry. If the completeness of the original basis set is not high, but too much diffusion function is added, the effect that should be manifested by the valence layer basis function will be manifested by the diffusion function, which may cause unreasonable results in the study of some problems. This actually belongs to the category of intramolecular BSSE. An example is that after the discovery of JACS, 128, 9342, the stable structure of benzene calculated by HF combined with some dispersive version of the Pople basis set (such as 6-31++G**) turned out to be curved, or a plane that should have been stable. The structure has false frequencies. This is because there is no higher angular momentum basis function (especially f) that the post-HF calculations rely on in these Pople basis sets. The more diffuse s and p basis functions are extended to fully demonstrate the higher angular momentum basis of carbon. The effect of the function causes the structure to bend.

**Common Basis Sets Containing Diffusion Function**

Let me talk about the general characteristics of the diffusion function. The exponent of the diffusion function of each angular momentum is several times smaller than the smallest exponent of the other equivalent angular momentum functions in the basis set. The number of diffusion functions and the angular momentum involved are different in various basis sets. The diffusion function is generally non-shrinking.

Since the diffusion function is important for many situations, most of the mainstream basis sets have versions with the diffusion function. Some versions with diffusion function were developed by the authors of the original basis group, and some were proposed by other researchers. Here are some common ones: Pople series basis set: only add a layer of sp (that is, a layer of s and a layer of p with the same exponent) to the diffusion function of the heavy atom and add a plus sign in front of G, such as 6-31+G *; If you also add a layer of s-diffusion function to hydrogen and helium atoms at the same time, add a plus sign in front of G, such as 6-311++G(2df,2p). There are many Pople series basis sets, but the exponents of the diffusion function are all shared and not optimized separately. (It is worth mentioning that the time-consuming of 6-311G* is lower than 6-31+G*. When the diffusion function is not critical, the result of the former is better than the latter. Countless people use 6- When 31+G*, I didn't know that 6-311G* is much more cost-effective for the problem they are studying!) Dunning related consistency basis set (cc-pVnZ series): The version with the diffusion function is aug-cc -pVnZ series (aug=augmented) is to add a layer of diffusion function of the same angular momentum to each angular momentum function of the corresponding cc-pVnZ basis set. For example, cc-pVTZ is 4s, 3p, 2d, 1f for C, so the aug- version will add a layer of s, a layer of p, a layer of d, and a layer of f diffusion function. And cc-pVDZ is 2s, 1p for hydrogen, so the aug-version will add a layer of s and a layer of p diffusion function to it.

The same diffusion function as aug-cc-pVnZ is added to the basis set of cc-pCVnZ and cc-pwCVnZ, which is related to the description of the nucleus, to become aug-cc-pCVnZ and aug-cc-pwCVnZ; added to cc-pVnZ suitable for DKH calculations -DK becomes aug-cc-pVnZ-DK; added to cc-pV(n+d)Z series (this basis set is based on cc-pVnZ by adding a layer of relatively tight d function to improve the extrapolation Convergence) becomes aug-cc-pV(n+d)Z, where only the exponent of the d diffusion function has been re-optimized. The relevant consistent pseudopotential basis set cc-pVnZ-PP also has a diffusion function version aug-cc-pVnZ-PP. The type and number of diffusion functions added are the same as the aug-cc-pVnZ series, but the exponents have been re-optimized. In addition, you can also add a multi-layer diffusion function to each angular momentum of the cc-pVnZ series. d-aug-cc-pVnZ and t-aug-cc-pVnZ add two and three layers of diffusion to each angular momentum. The function is extremely expensive, and is generally used for the purpose of accurately calculating Rydberg excited state, hyperpolarizability, etc. Ahlrichs' def2-series basis sets: There is currently no official version of this type of basis set with diffusion function. Jensen's polarization consistency basis set pc-n [13]: Jensen proposed a method of adding a diffusion function to his pc-n series basis sets. The high angular momentum of the diffusion function can be selected according to needs. The results show that the addition of s and p diffusion can greatly improve the accuracy of the electron affinity calculated by DFT, and the calculation results of further improving the response properties also need to add a diffusion function with higher angular momentum. Lanl pseudopotential basis set [14], the author added a layer of d-polarization and a layer of p-diffusion function to the main family Lanl2DZ, named LANL2DZdp, in terms of calculating electron affinity, vibration frequency, and bond length Both are greatly improved than Lanl2DZ. LANL2TZ+ and LANL08+ add a layer of diffusion to the first-period transition metal for LANL2TZ and LANL08, respectively. This is because the first-period transition metal filled with the d shell is sometimes easily polarized. Except for the exponents of the diffusion function of the basis sets such as LANL2TZ+ and LANL08+, which are derived using the even-tempered method, most of the exponents of the diffusion function mentioned above are derived from different methods to minimize the energy calculated by the anion (Note: the original basis The contraction coefficient and exponent of the group remain unchanged, and the diffusion function is simply added to the original basis group). However, the basis set of diffusion function with good calculation energy is used to calculate other attributes, especially the (hyper)polarizability that strongly depends on the diffusion function may not be good, or the cost performance is not high. In order to make the response properties such as (hyper)polarizability have satisfactory calculation results at a lower computational cost, many people make the diffusion function and even the entire basis set directly derived from the calculation of the optimized response properties. A few examples:

Sadlej POL: Also called the Sadlej pVTZ basis set. It has been proposed since 1988. The parameters are obtained by optimizing the calculation of the polarization rate. The size is close to cc-pVTZ, and the accuracy of calculating the polarization rate is close to the much more expensive aug-cc-pVTZ.

Sadlej ZPOL: Proposed successively in 2004. Simplify the POL basis set. It is suitable for calculating the dipole moment and polarizability of very large systems, which is as time-consuming as 6-311+G*, but the accuracy of calculating the polarizability is better than 6-311++G (2df, 2p).

Sadlej LPol-ds: Proposed in 2009. The base component of LPol series is LPol-ds/dl/fs/fl, which increases sequentially. LPol-ds is the smallest among them, but it is also much larger than POL. The accuracy of calculating the first hyperpolarization rate is excellent, similar to d-aug-cc-pVTZ, and the time-consuming is the lowest in the same grade of accuracy. There are only definitions for C, H, O, N, and F.

def2-SVPD, TZVPD, TZVPPD, QZVPD, QZVPPD basis set: proposed in JCP, 133, 134105, is to add a diffusion function to the def2- series of SVP, TZVP, TZVPP, QZVP, QZVPP basis sets. The exponent of the diffusion function is obtained by optimizing the HF polarizability of the atom.

LFK pseudopotential basis set: It adds diffusion and polarization functions to 39 main group elements on the basis of SBKJC pseudopotential basis set. The purpose is to achieve the polarization rate under the all-electron Sadlej basis set under the calculation of pseudopotential basis set calculation accuracy. Exponents and shrinkage coefficients are obtained by approximating experimental atomic polarizabilities or high-precision relativity calculations.

If no one has proposed a version with a diffusion function for a certain basis set, but the calculation requires a diffusion function, then you can consider using the diffusion function of other basis sets of similar size. It can also be constructed based on the existing basis set in an even-tempered way. In addition, some people have proposed that the exponents of s and p with the smallest exponents in the basis set can be divided by 3 as the exponent of the s and p diffusion function [15]. Although the index calculated in this way is certainly not as good as an index specifically optimized for a specific problem, but for a wider range of systems and types of problems that are not considered in the process of optimizing the index, the directly calculated index may also perform better.

# Chapter 3
# Calculation and Analysis of Electron Transition Spectra

## 3.1 Calculation Method of Excited States

### 3.1.1 Introduction

The theoretical calculation of the excited state is usually these:

(1) Vertical excitation energy: Vertical excitation is purely an excitation process imagined by theorists, that is, the structure does not change during the transition process. Its calculation is the easiest, and its value corresponds approximately to the maximum absorption peak of the experiment (because vibration coupling is not considered, or the actual process is not so vertical, so even if the vertical excitation energy calculation is absolutely accurate, it will often be high Estimating the maximum absorption peak position 0.1–0.3 eV) is very useful for identifying and predicting spectra. Therefore, most studies on excited states are calculated as vertical excitation energy. Fluorescence is often counted. The vertical emission energy is usually calculated under the excited state equilibrium structure. The principle is very similar.

(2) Adiabatic excitation energy: Adiabatic excitation is also purely an excitation process assumed by theorists, that is, the excitation process transitions from the minimum point structure of the initial state potential energy surface to the minimum point structure of the final state potential energy surface. The calculation of adiabatic excitation energy requires excited state optimization, which is time-consuming, and is usually not as close to the maximum absorption peak as the vertical excitation energy, so there are relatively few studies. However, the adiabatic excitation energy can be used as a good approximation of the experimentally obtained 0–0 transition energy (but because the excited state ZPE is usually lower than the ground state, the accurately calculated adiabatic excitation energy tends to overestimate the actual 0–0 value by about 0.1 eV).

(3) Oscillator intensity: The intensity of the oscillator of each state plus the excitation energy of each state, after broadening, the theoretically calculated spectrum can

© Tsinghua University Press 2023

M. Sun and X. Mu, *Computational Simulation in Nanophotonics and Spectroscopy*,
Nanoscience and Nanotechnology,
https://doi.org/10.1007/978-981-99-4732-4_3

be obtained. Obviously, the intensity of the oscillator is a very important quantity. If you don't know how the spectrum is obtained based on the theoretical calculation data. But note that in order to accurately obtain the absorption/emission spectrum, including the prediction of the color of the dye molecule, the entire absorption band must be predicted more accurately, not just focusing on a few absorption peaks, which must consider vibration For the coupling of dynamics and electronic states.

(4) Excited state geometric structure: What kind of change occurs in the geometric structure of the system caused by electronic excitation is a question of chemistry of interest. Sometimes the changes are significant, such as the transfer of hydrogen and the large-scale movement of molecular structures.

(5) Potential energy surface: In fact, the previous study is only a few points on the potential energy surface of the excited state, and many studies require more information about the potential energy surface of the excited state. Including the investigation of the potential barrier on the potential energy surface of the excited state, the study of the vibration mode of the excited state, the cone/avoidance of intersection with the ground state, etc. These issues are extremely important for the study of photochemical reactions and non-radiative transitions (intersystem crossing, internal conversion) Crucially, this is also relatively advanced research.

(6) The electronic structure properties of the excited state: such as the dipole moment, quadrupole moment, polarizability, atomic charge, bonding mode, aromaticity, etc. of the excited state, which requires the wave function of the excited state.

### 3.1.2   TDDFT

In the early years, people often calculated the excited state at the semi-experience + CI level, which was very rough. Later, the standard method was to do CIS based on HF orbit, but the accuracy of the excitation energy of CIS was quite unsatisfactory. Later, TDDFT became popular and became a standard method for ordinary people to calculate excited states. It took no more time than CI, and the accuracy was significantly improved. However, early TDDFT did not support analytical gradients of excited states (such as g03), and numerical gradients were needed. Therefore, optimizing the structure of excited states is time-consuming. Each step of optimizing a system with N geometric variables takes time equivalent to 6N excitation energy calculations. Therefore, during that period, CIS was popular to optimize the excited state structure and TDDFT was used to calculate the excitation energy. However, mainstream quantization programs, including g09, have begun to support TDDFT analytical gradients, and CIS is basically useless. However, the analytical second derivative of TDDFT has not yet been popularized. Only a few programs such as Q-Chem can do it. Therefore, it is necessary to calculate the excited state frequency, especially the calculation of the vibration coupling closely related to it, which is still extremely important for larger systems. Time-consuming, another option is to make do with CIS that supports second-order analytic derivatives.

In short, TDDFT is undoubtedly the most popular method of calculating excited states. There are two key points that affect the accuracy of TDDFT, one is the choice of functional, and the other is the choice of basis set. The basis set is actually not that arbitrary. For general valence layer excitation, 6-31G* is the bottom limit (suitable for larger systems). If the conditions are better, use 6-311G*. If there is enough calculation, you can use TZVP. If you can use more advanced def2- TZVP is perfect in terms of the basis set. If it involves higher-order valence layer excitation, especially Rydberg excited state, or the calculated anion system, the dispersion function is indispensable, and it cannot be lower than 6-31+G*, generally use aug-cc-pVTZ. If you want to calculate the Rydberg excitation more accurately, it is best to use the d-aug-cc-pVTZ basis set, which adds a layer of dispersion function than aug-cc-pVTZ, but the accuracy of TDDFT cannot match this basis set. Accuracy, if you want to use it, you should change the method to a more accurate method.

Compared with the choice of basis set, the choice of functional is larger and there are more doorways. Let's briefly talk about the currently recognized better choice. The previous section mentioned that the study of excited states will involve a lot of content, and how to measure the calculation accuracy of excited states is a question. Testing the vertical excitation energy is obviously the simplest and direct method, but the vertical excitation energy cannot be measured experimentally. Considering the vibration coupling to compare the maximum absorption peak is too time-consuming (to calculate the second derivative), so theoretical chemists usually use high precision Methods such as CC3, CASPT2, MRCI combined with large basis sets to calculate the vertical excitation energy are used as the gold standard to measure the rationality of other cheap methods. The errors mentioned below are mostly relative to the gold standard given by such theoretical methods. (But through theoretical methods to obtain the difference between the vertical and 0–0 excitation energy, and the difference between the gas phase and the solvent phase excitation energy to correct the experimental experiment 0–0 data, it can also estimate the experimental vertical excitation energy)

When the valence layer is locally excited and the excited state is a singlet state: PBE0 is the best choice, and the error is about 0.1–0.3 eV. B3LYP is second only to PBE0, and the error is slightly higher than PBE0 from a statistical point of view. There is a trend worth mentioning, that is, the larger the HF component, the larger the calculated excitation energy value. Pure functionals have no HF part and the excitation energy is obviously low, while those with higher HF components, such as BHandHLYP (50%) and M06-2X (54%), tend to overestimate the excitation energy. The range separation functionals mentioned below, such as CAM-B3LYP, wB97XD, LC-wPBE, etc., have high excitation energy due to the extremely high remote HF component. As for the functionals of HF components such as PBE0 25% and B3LYP 20%, from the average point of view, there is no obvious overall tendency to overestimate or underestimate the excitation energy. The above rules are applicable to most systems, but for large conjugated systems (such as dye molecules), some benchmark articles [16] found that 25% is low, and there is a tendency to underestimate the excitation energy, and it is recommended to use Higher HF content such as PBE38

(37.5%) or BMK (42%). The HF components of common functionals are summarized in the appendix: List of HF components of different DFT functionals.

The valence layer is locally excited, and the excited state is a triplet state: almost all methods underestimate the excitation energy regardless of the HF composition. Although the influence trend of HF component on excitation energy still exists, it is far less obvious than the singlet state. M06-2X performed well in this situation, while PBE0/B3LYP performed mediocre.

Charge transfer excitation and Rydberg excitation: charge transfer excitation means that electron excitation will cause obvious electron-hole separation, leading to long-range electron density transfer; Rydberg excitation means that electrons are excited to a higher order and show obvious dispersion Excitation on the characteristic empty orbit. Because the long-range behavior of traditional DFT exchange functionals is wrong, the errors in calculating these two types of excitation energies are relatively large, unless there is a high HF component like M06-2X. After 2000, a class of range separation functionals began to appear. The more representative ones include CAM-B3LYP, wB97XD, LC-wPBE, etc., which use very high or even 100% HF components for the long-range part, which makes these two types of exciting The error is greatly reduced, but, as a price, the calculation accuracy of the local excitation energy is not small compared to PBE0. For charge transfer and Rydberg excitation, I recommend CAM-B3LYP, wB97XD, and M06-2X. There is no clear determinant which is better.

If you don't want to distinguish between various types of excitations, and just want to use a functional to calculate, then I recommend M06-2X or wB97XD, because they are not necessarily ideal for all types of excitations, but they are certainly not scumbags, and the overall performance is better. it is good. As for CAM-B3LYP, although the charge transfer excitation and Rydberg excitation perform well, the local excitation is too bad. Both M06-2X and wB97XD were born in the same year (2008). They are very promising functionals. The two key weaknesses of traditional functionals (charge transfer, Rydberg excitation failure, and poor description of weak interaction) They are all solved, and other aspects are good, but the disadvantage of B3LYP compared to the old comrades is that their calculations are relatively time-consuming, especially wB97XD, which often takes twice or several times longer. By the way, wB97XD was born following wB97X. It not only adds dispersion correction on the basis of wB97X, but also adjusts other parameters. The calculated excitation energy of wB97XD has a slight advantage over wB97X, and at the same time the dispersion correction is taken into consideration (without additional time-consuming), so wB97XD is recommended.

In particular, it is worth mentioning that the w-regulated functional has been particularly popular in recent years. The calculation result of the range separation functional significantly depends on the w parameter, just as the result of the ordinary functional significantly depends on the hybrid component. The w parameter is different for different systems. Therefore, the w parameter must be optimized before calculating a system using the w-regulated functional (also called w control literally). The accuracy of the excitation energy calculation using the functional adjusted by w is quite good, and it is almost always better than the empirically selected functional, but unfortunately the process of w optimization is more time-consuming.

The traditional low-HF functional calculation charge transfer excitation is bad, which is not only reflected in the excitation energy, but also in the structure. For example, for a molecule in J. Chem. Theory Comput., 2012, 8, 2359–2372, the two adjacent benzene rings in the excited state structure of CAM-B3LYP are nearly parallel, while the optimized B3LYP is perpendicular to each other. of. Therefore, when investigating the charge transfer system, ordinary hybrid functionals with low HF content should not be used in optimization, and pure functionals are absolutely not used.

Never use GGA functionals for TDDFT calculations. Not only do you seriously underestimate the excitation energy, but because of the self-interaction error (SIE), so-called ghost states are prone to occur for large conjugate systems. These ghost states have very low excitation energy, have the characteristics of charge transfer, and the oscillator strength is very small, which is a false state with no physical meaning. Functionals with low HF components like B3LYP may also appear ghost states. Although the intensity of the ghost state oscillator is small, it does not affect the obtained absorption spectrum much, but it will take more calculation time to make the truly meaningful states with higher excitation energy not counted.

The double hybrid functional represented by B2PLYP is also used in TDDFT calculation, which is actually similar to CIS(D) calculation, that is, perturbation correction is made on the basis of TDDFT. The amount of calculation is obviously larger than that of ordinary TDDFT, but the accuracy is also improved accordingly. For singlet excited states in the valence layer, the error can be reduced by 30% compared to PBE0. At present, only ORCA supports double-hybrid functionals for TDDFT calculation in the form of CIS(D). B2PLYP is considered poor in the double hybrid functional. As shown in JCP, 132,184103, it will be better to use B2GP-PLYP for TDDFT.

The solvent must be considered in the calculation of the excited state, and the solvent effect has a greater influence on the excited state than on the ground state. In particular, the influence of the solvent on the spectrum is obviously not negligible. Due to the different interactions between the ground state and the excited state of the solvent and the solute, the energy level changes to different degrees, so the solvent will redshift or blue shift the absorption peak. In terms of solvents, TDDFT can already be handled relatively well, and can be perfectly combined with mainstream PCM models. Non-equilibrium solvent effects and state-specific calculations can be considered to respond to the excited state density. The explicit solvent model can also be used when it is necessary to describe the strong interaction with the solute. Sometimes the actual external environment is more complicated, such as a protein environment. At this time, this effect can be expressed by QMMM or simpler background charges. However, effects such as spectral peak broadening and disappearance of spectral details due to solvents cannot be represented by implicit or explicit solvent models. Simulations are required, which is very complicated.

After continuous improvement, TDDFT seems to be pretty good now. What weaknesses are left? The author believes that there are the following two points, which are restricted by the form of TDDFT and are difficult to deal with: (1) It is not suitable to study the potential energy surface intersection, which is still the world of multi-reference methods.

(2) Not suitable for studying excited states with obvious two-electron excitation characteristics (although it is better than CIS).

The rest to continue to improve is (1) popularize the second-order analytic derivative of TDDFT as soon as possible (2) further improve the functional to increase the accuracy, and consider optimizing the functional specifically for calculating the excitation energy.

A problem often encountered when calculating absorption spectra is to determine the number of states to be calculated. Often the 2–7 eV region is the region of interest in the UV–Vis spectrum. If the number of states is calculated less, the high-energy excited states will be missed, and you have to make up the calculation (for example, if you calculate 20 states first, if you find that it is not enough, then make up 10 states, which will cost a lot more than 30 states directly. Time); however, the more states the TDDFT takes, the longer it takes, so it cannot be counted too much at once. How many states need experience is closely related to atomic number. When the number of atoms is small, the highest excitation energy may be high enough when the number of atoms is 10, but when the number of atoms is large, such as an organic molecule with 50 atoms, it may take nearly 100 states to fully cover the absorption spectrum of interest. area. In addition to judging how many states need to be counted through experience, you can also use the sTDA method proposed by JCP,138,244104 to do a TDDFT approximate calculation. This method requires almost no additional time-consuming relative to the ground state single-point calculation. See approximately how many states need to be calculated to achieve the desired excitation energy, and then do an accurate TDDFT calculation. If the system is too large and the number of states to be calculated is too many, for example, there are hundreds of states, then simply use the result of sTDA directly, although the deviation of the UV–Vis spectrum calculated by accurate TDDFT is not small.

### 3.1.3  Other Calculation Methods Excited States

This section will take a random inventory of methods for calculating excited states other than TDDFT.

CIS: The most classic method of calculating excited states known to women and children. The theory is very clear, and it is also the basis of most other methods of calculating excited states. It has size consistency. The accuracy of this method is quite unsatisfactory, especially when there is a tendency to seriously overestimate the excitation energy, often up to 1–2 eV, so some people multiply it by a correction factor to avoid the overestimation trend (but not universal). The reason for this trend is that its orbits are derived from ground state HF calculations (the single-excitation configuration function generated is not ideal for CIS variation, and the energy is high), and the dual excitation configuration is not introduced. Describe the dynamic correlation of excited states to reduce their energy. The advantage of CIS is that the calculation is faster, the calculation of charge transfer excitation is better, and there are second-order analytical derivatives in Gaussian. In short, CIS is also used for the purpose of qualitative calculation. Generally, it can give the correct order of excited state energy, but do not expect quantitative and accurate calculation.

CISD: This method generally calculates the ground state, but it can also calculate the ground state and the excited state at the same time. The excitation energy is the difference between them. However, the dynamic correlation of the ground state in CISD is much more fully considered than the excited state. Therefore, when calculating the excitation energy, the error offset is not good, and the excitation energy will be overestimated. It is not as good as the cheaper CIS(D), and the amount of calculation is large. No one uses it at all.

Full CI: The Holy Grail!

Coupling clusters: Equation of Motion (EOM) and Linear Response (LR) are processing methods that use coupled clusters (CC) to calculate excited states. The former is based on non-time-dependent processing, and the latter is based on time-dependent processing, but the final result is the same, Can be combined with CCSD, CCSDT..., LR can also be easily combined with the iterative approximation versions of the coupling clusters of each order, such as CC2/3, CCSDT-1/2/3, to calculate the excitation energy, which is the same size as the ground state CC Consistent, the scale and the ground state CC are also consistent. The following specifically talk about several commonly used. EOM-CCSD: It is widely used, Gaussian has. The results of the single reference state system are good, and the single electron excitation error is less than 0.3 eV, which is stronger than TDDFT. It is not good for the multi-reference state system, and the system error with obvious two-electron excitation characteristics is also large. Slightly expensive, the system size that can be calculated is much smaller than the size that TDDFT can handle. EOM-CCSD(T): Non-iterative CCSDT approximate versions such as CCSD(T) can also be combined with EOM/LR to calculate the excited state, but the specific implementation method is very complicated, not as clear as the CCSD(T) calculation ground state, so it is not popular, But the accuracy is quite good, especially for the double electron excitation accuracy. EOM-CCSDT: It is very expensive and can only be used for a few atoms. Its approximate version EOM-CCSDT-3 is still somewhat less accurate than EOM-CCSDT, but it is still significantly improved over EOM-CCSD, especially for dual-electron excitation. Generally, the error of both single and double excitation is less than 0.05 eV. EOM-CCSDTQ can only be used in the model system, the error is less than 0.02 eV, and it is in good agreement with FCI. It can be done with NWChem. EOM-IP-CCSD, EOM-EA-CCSD: For the dual state, the usual EOM-CCSD is based on UHF, and the result may not be good. If the system becomes a closed shell after adding or subtracting one electron, it is suitable to use EOM-IP-CCSD and EOM-EA-CCSD to calculate the spectrum.

LR-CC3: CC3 is an approximate version of CCSDT, the accuracy is between CCSD and CCSDT, >=CCSD(T) level, but it is much more expensive than CCSD(T). Combined with LR, the excited state can be calculated, which is slightly weaker than EOM-CCSDT, but the accuracy is already excellent. CCSD(T) is regarded as the gold standard for weak interaction calculations, while LR-CC3 can be regarded as the gold standard for excitation energy calculations, which can be safely used as a reference value. However, when the multi-reference state feature is strong, or the two-electron excitation feature is obvious, CASPT2 is more suitable. Dalton is quite suitable for LR-CC3 excitation energy calculation.

MCSCF: Although it can be used to calculate excitation energy, the main purpose is not to quantitatively and accurately calculate excitation energy, but to provide perturbation, CI and other methods that consider dynamic correlations with reference state wave functions that fully consider static correlations. Taking the dynamic correlation into account can get a very accurate excitation energy. In addition, only the use of MCSCF or methods based on it can accurately study the problem of potential energy surface intersection, because only MCSCF can describe the two intersecting states in a balanced manner (if it is only S0-T1 intersection, DFT can also be used). MCSCF includes CASSCF and RASSCF, and CASSCF is generally used. Although RASSCF seems to be more free and flexible in setting, it is possible to do more with less, it uses RAS1 and RAS3 space to enclose a larger active space than CASSCF in a cheap and discounted way (limiting the number of excited electrons). However, it is nondescript, the description of static correlation is not thorough, and the description of dynamic correlation is half-hearted, so it is not widely used.

CASPT2: This is the most commonly used method for high-precision calculation of excitation energy of small molecules. The accuracy is good, the error is less than 0.2 eV, and the calculation amount is not very outrageous. The accuracy is slightly inferior to CC3 but the calculation amount is much lower. In particular, CASPT2 is more suitable than CC3 for the excited state with strong multi-reference characteristics of the ground state and obvious two-electron excitation characteristics. CASPT2 is the trump card for studying photochemical problems (involving the intersection of various potential energy surfaces). But the key disadvantage is that it requires experience in the selection of active space, and the result is not good if the selection is not good. It is not like a black box like a coupled cluster, and the ease of use is therefore compromised. In addition, there are sometimes intrusive state problems (through energy level movement. solve). Note that CASPT2 is specifically divided into two types: single-state version SA-CASPT2 and polymorphic version MS-CASPT2. Excited state calculations generally use the latter. The amount of calculation is greater than SA-CASPT2, but it can solve SA-CASPT2 for potential energy surface crossing and valence layer- Possible problems in Rydberg mixed state. Molpro and Molcas are the main force for CASPT2, which can give the vibrator intensity and therefore predict the spectrum. Molpro also supports the analytical gradient of CASPT2, so it can easily optimize the excited state. In addition, Molpro also has CASPT3, which is more accurate but more computationally intensive and uses less. There are many other methods for second-order perturbation based on MCSCF similar to CASPT2, and some are also suitable for calculating excitation energy, such as XMCQDPT2 supported by Firefly, QD-NEVPT2 supported by ORCA, etc. The popularity is far less than that of CASPT2, not mentioned here.

MRCI: Do CI based on MCSCF wave function, generally MRCISD is used. The amount of calculation is quite large, significantly higher than CASPT2, so only very small molecules can be studied. The accuracy is correspondingly close to the FCI limit, which is more accurate and more reliable than CASPT2.

SORCI (Spectroscopy Oriented CI): MRCI-based method, which also introduces multi-reference second-order perturbation, which is specially suitable for the study of excitation energy, and strives to achieve high accuracy and good calculation efficiency. The amount of calculation is less than MRCI and can be used in a slightly larger system.

SAC-CI (SAC-Configuration Interaction): EOM-CC and LR-CC are theoretically equivalent to it, but the algorithm is different. SAC-CI can be said to be an approximation of EOM-CCSD. Because it is not widely supported by the quantization program (although Gaussian supports it), the advantages are not obvious and it is not widely popularized.

DFT/MRCI: Grimme proposed in 1999 that the idea is to build a CI matrix based on DFT orbits and exchange correlation potentials with DFT to make MRCI. In order to avoid dynamically related double-counting, the non-diagonal matrix elements of the CI matrix are scaled to exponentially reduce to 0. And because correlation functionals have considered a lot of dynamic correlation, only important configuration functions are selected to consider dynamic correlation. DFT/MRCI saves a lot of time than traditional MRCI, and can also introduce RI approximation to further reduce time consumption. The calculated excitation energy is better than TDDFT, the triplet excitation energy error is twice lower than the best ordinary functional, and the singlet excited state has a smaller advantage. However, DFT/MRCI is not supported in mainstream quantitative programs. The program can only be requested by the author. Therefore, although the proposal is not short, it does not reflect the practicality.

Various semi-empirical methods: The method of CIS to calculate excitation energy based on various semi-empirical methods was used more in the early days, but is rarely used now, that is, ZINDO, which optimizes parameters for excitation energy calculations, is calculating It is still often used in large systems. Among those ordinary semi-empirical methods, the OM3 semi-empirical method developed by Walter Thiel's group may be the best, with an error of twice that of TDDFT. However, since the program for doing OM3 calculations is not public, it is obviously not used by anyone.

ZINDO: ZINDO is divided into ZINDO/1 and ZINDO/S. The former is used to calculate the ground state, and here is the latter, also called INDO/S. It was developed based on INDO in the 1970s by Zerner et al. It is a semi-empirical method that is very suitable for calculating excitation energy. The results for organic molecules are very good, better than CIS, and especially suitable for pi-pi* and n-pi* excitation. ZINDO can also be regarded as a lot of transition metals in addition to the main group. The disadvantage is that it is not suitable for systems that contain electron transfer processes, higher-order excited states, Rydberg states, anionic systems, metals with unpaired electrons, and two-electron excitation characteristics that cannot be ignored. And it is not suitable for calculations of excited states other than excitation energy, such as excited state optimization, potential energy surface intersection, etc. Since ZINDO is an optimized parameter for calculating the spectrum, calculating the ground state energy is not as good as AM1 and PM3. ZINDO is generally done at the CIS level. If dual excitation (CISD) is considered, the ZINDO results can be further improved (especially to improve the accuracy of higher-order excitation and vibrator intensity). Zerner's original ZINDO program, Gaussian, Arguslab, Cache, Hyperchem all support ZINDO.

INDO/X (INDO for eXcited states): It was released in 2014. It is the re-parameterization of ZINDO. The training set mainly uses the data of the TBE-2 vertical excitation energy test set. The triplet and singlet states are changed to the same set of Hamilton (ZINDO is two different sets). The average error is twice as small as that of ZINDO, and the accuracy of the vibrator intensity is very good. From the statistical data, it can reach the TDDFT level, but the error for some excited states of some systems is still large, not as robust as TDDFT.

### 3.1.4  Appendix: List of HF Components of Different DFT Functionals

Here, the HF exchange components of common hybrid functionals are listed in order from low to high. Some types of calculations directly depend on the HF content, so it is useful to have such a list. For example, it is well known that the excitation energy is directly dependent on the HF component. GGA is too red and HF is too blue. The higher the HF component, the higher the hybrid functional excitation energy. Therefore, if you find that there is a systematic deviation between the TDDFT calculation and the experimental spectrum, you can choose a functional with an appropriate HF ratio to use (although some people adjust the HF composition by themselves to make the experiment just right, but this is easy to be questioned by the reviewers. People feel that they are deliberately trying to make up, and there is a suspicion of falsification. If you use the ready-made functionals, there is no such trouble). The ones marked in bold are worthy of main attention, covering several major files (see Table 3.1).

## 3.2  Analysis Method of Excited States

### 3.2.1  Hole-Electron Analysis

Hole-electron (hole-electron) analysis is a very powerful and extremely practical method for investigating the characteristics of electron excitation. It describes the electron excitation process as "hole → electron", which can be used graphically. Intuitively examine where the electrons leave and go, whether they are local excitation, global excitation, electron transfer excitation, or hybrid feature excitation. It can also quantitatively investigate the electron transfer distance, the degree of separation between holes and electrons, the contribution of atoms or fragments to the excitation

**Table 3.1** List of HF components of different DFT functional

| Functional | HF (%) |
|---|---|
| GA, meta-GGA | 0 |
| TPSSh | 10 |
| O3LYP | 11.61 |
| TPSS1KCIS | 13 |
| MPW1KCIS | 15 |
| Hybrid version B97 | 19.43 |
| B3LYP, B3P86, B3PW91 | 20 |
| B97-1, B97-2, HCTH93 | 21 |
| MPW3LYP, X3LYP | 21.80 |
| PBE1KCIS | 22 |
| APFD | 23 |
| PBE0, B1B95, TPSS0 and mPW1PW91 | 25 |
| M06 | 27 |
| PW6B95, M05 | 28 |
| MPW1B95 | 31 |
| PBE0-1/3 | 33.33 |
| PBE38 | 37.50 |
| BB1K, BMK | 42 |
| MPW1K | 42.80 |
| MPWB1K | 44 |
| MN15 | 44 |
| PWB6K | 46 |
| BHandHLYP, PBE50 | 50 |
| M08-HX | 52.23 |
| M06-2X | 54 |
| M05-2X | 56 |
| M08-SO | 56.79 |
| M06-HF | 100 |

of electrons, the Coulomb attraction energy of holes and electrons, and so on. This analysis can be used for electronic excitation tasks done by CIS, TDHF, TDA-DFT and TDDFT methods. Definition of holes and electrons in the process of electron excitation For CIS, TDHF, TDA-DFT and TDDFT, the excited state wave function is described by a linear combination of single excitation configuration functions. For CIS and TDA-DFT, each configuration function has a coefficient $\omega$; for TDHF and TDDFT, each configuration function can be used as an excitation configuration and has a coefficient $\omega$, or as a de-excitation Configure and have the coefficient $\omega'$, After derivation, for TDHF/TDDFT, the expression of holes and electrons can be written in the following form.

$$\rho^{\text{hole}}(\mathbf{r}) = \rho^{\text{hole}}_{(\text{loc})}(\mathbf{r}) + \rho^{\text{hole}}_{(\text{cross})}(\mathbf{r})$$

$$\rho^{\text{hole}}_{(\text{loc})}(\mathbf{r}) = \sum_{i \to a} \left(w_i^a\right)^2 \varphi_i \varphi_i - \sum_{i \leftarrow a} \left(w_i^{\prime a}\right)^2 \varphi_i \varphi_i$$

$$\rho^{\text{hole}}_{(\text{cross})}(\mathbf{r}) = \sum_{i \to a}\sum_{j \neq i \to a} w_i^a w_j^a \varphi_i \varphi_j - \sum_{i \leftarrow a}\sum_{j \neq i \leftarrow a} w_i^a w_j^{\prime a} \varphi_i \varphi_j$$

$$\rho^{\text{ele}}(\mathbf{r}) = \rho^{\text{ele}}_{(\text{loc})}(\mathbf{r}) + \rho^{\text{ele}}_{(\text{cross})}(\mathbf{r}) \tag{3.1}$$

$$\rho^{\text{ele}}_{(\text{loc})}(\mathbf{r}) = \sum_{i \to a} \left(w_i^a\right)^2 \varphi_a \varphi_a - \sum_{i \leftarrow a} \left(w_i^{\prime a}\right)^2 \varphi_a \varphi_a$$

$$\rho^{\text{ele}}_{(\text{cross})}(\mathbf{r}) = \sum_{i \to a}\sum_{i \to b \neq a} w_i^a w_i^b \varphi_a \varphi_b - \sum_{i \leftarrow a}\sum_{i \leftarrow b} w_i^{\prime a} w_i^b \varphi_a \varphi_b$$

where the r is the coordinate vector, $\phi$ is the orbital wave function, i or j is the occupied orbital label, and a or b is the empty orbital label. Therefore, for example, $\sum_{i \to a}$ represents looping each excitation configuration, and $\sum_{i \leftarrow a}$ represents looping each excitation configuration. Both the hole distribution $\rho_{hole}$ and the electron distribution $\rho_{ele}$ are divided into two parts: local term and cross term. The local term generally dominates and reflects the contribution of the configuration function itself, and the cross term cannot be ignored, otherwise the quantification is inaccurate. It reflects the influence of the coupling between the configuration function on the distribution of holes and electrons. In fact, if an electron excitation can be perfectly described by a pair of orbital transitions $\sum_{i \to a}$, in other words the contribution of this configuration function is exactly 100%, then $\phi_i$ and $\phi_a$ can be Directly ideally as holes and electrons respectively. The holes and electrons defined in the above way are equivalent to $\phi_i^2$ and $\phi_a^2$ respectively, which can be regarded as the electron density corresponding to the i and a orbitals (assuming that these two orbitals are both single-Happening). Therefore, the electrons and holes defined above have no phase information because the orbital wave function is squared. Since the actually calculated electronic excited state is always participated by many configuration functions, no pair of orbital transitions can describe the electronic excitation perfectly. Therefore, in order to fully describe the electrons and holes in this situation, Dr. Lu created Based on the above-mentioned specific definitions of electrons and holes, it takes all orbital transitions into consideration, which can ideally, comprehensively and fully demonstrate the characteristics of electronic excitation [17].

### 3.2.2 Quantitative Description

In order to facilitate the measurement and discussion of electronic excitation characteristics through some quantitative values, Dr. Lu defined the Sm and Sr indices, that is, the Sm and Sr functions are integrated in the whole space:

$$S_m \text{ index} = \int S_m(\mathbf{r})d\mathbf{r} \equiv \int \min\left[\rho^{\text{hole}}(\mathbf{r}), \rho^{\text{ele}}(\mathbf{r})\right]d\mathbf{r}$$
$$S_r \text{ index} = \int S_r(\mathbf{r})d\mathbf{r} \equiv \int \sqrt{\rho^{\text{hole}}(\mathbf{r})\rho^{\text{ele}(\mathbf{r})}}d\mathbf{r} \tag{3.2}$$

The larger the two indexes, the higher the degree of overlap between holes and electrons; the smaller the value, the more significant the separation of holes and electrons. The Sr index must be greater than the Sm index. The range of these two indices is both [0,1], 1 means that holes and electrons overlap perfectly, and 0 means that there is no overlap at all.

Bahers et al. proposed a series of indicators based on the density difference to analyze the electronic excitation characteristics [18]. First of all, $\sigma$ can be defined for both holes and electrons. Its x, y, z three components are equivalent to the root mean square deviation (RMSD) of the hole or electron distribution in the x, y, and z directions, which reflects the distribution or dispersion degree. For example, the x component of $\sigma$ is defined as:

$$\sigma_{\text{hole},x} = \sqrt{\int (x - X_{\text{hole}})^2 \, \rho^{\text{hole}}(\mathbf{r})d\mathbf{r}} \tag{3.3}$$

Then the following quantities can be defined:

$$\Delta\sigma_\lambda = \sigma_{\text{ele},\lambda} - \sigma_{\text{hole},\lambda} \quad \lambda = \{x, y, z\}$$
$$\Delta\sigma \text{ index} = |\sigma_{\text{ele}}| - |\sigma_{\text{hole}}|$$
$$H_\lambda = \left(\sigma_{\text{ele},\lambda} + \sigma_{\text{hole},\lambda}\right)/2 \quad \lambda = \{x, y, z\}$$
$$H_{CT} = |\mathbf{H} \cdot \mathbf{u}_{CT}| \tag{3.4}$$
$$H \text{ index} = \left(|\sigma_{\text{ele}}| + |\sigma_{\text{hole}}|\right)/2$$
$$t \text{ index} = D \text{ index} - H_{CT}$$

Here are the uses of these quantities in turn

$\Delta\sigma_\lambda$: measure the difference in the spatial distribution of electrons and holes in the direction of $\lambda$.

$\Delta\sigma$ index: reflects the difference in the overall spatial distribution of electrons and holes.

$H_\lambda$: Measures the average extent of holes and electrons in the direction of $\lambda$.

$H_{CT}$: Measures the average extent of holes and electrons in the CT direction. The bold H in the formula is the vector written together by $H_x$, $H_y$, and $H_z$. The bold $\mu_{CT}$ is the unit vector in the CT direction, using the centroid positions of electrons and holes Available directly.

H index: reflects the overall average distribution breadth of electrons and holes

t index: a measure of the degree of separation of holes and electrons

The t index greater than 0 implies that the separation of holes and electrons is sufficient due to CT, because the centroids of holes and electrons are far away, and their average degree of extension in this direction is relatively not so high. If t index less than 0, it can be considered that there is no significant separation of holes and electrons in the CT direction, because the centroid distance between holes and electrons is not so large relative to their average extension.The common classification of electronic excitation is LE (local excitation), CT (charge transfer excitation), and Rydberg excitation. Using the above index, the types of electronic excitation can generally be distinguished as follows. Local excitation (or overall excitation: D index is small, Sr is large, t index is obviously negative, and the $\Delta\sigma$ index is not large. The main distribution ranges of holes and electrons in this excitation are very close, and the degree of overlap is obviously also It can't be lower. The distribution of holes and electrons is not significantly separated, and the distribution width is the same. Unidirectional charge transfer excitation: D index is large, other indicators are not necessarily large. Since it is CT excitation, the distance between electrons and holes must be large, and holes The degree of overlap with electrons can be large or small. When the overlap is large, it means that the separation of electrons and holes is not sufficient, and when the overlap is small, it means that the distribution of electrons and holes has been highly separated. Centrosymmetric charge transfer excitation: some When the system electrons are excited, the charge transfer can be carried out in multiple directions. The center-symmetric charge transfer excitation is an ideal situation. It should have a small D index and a large $\Delta\sigma$ index. Rydberg excitation: D And Sr index are not large, the $\Delta\sigma$ index is very large. The difference between the hole and the centroid distance of this kind of excitation is generally not significant, but the electron distribution range is far more diffuse than the hole distribution range, so the degree of overlap will not be too high The above judgment method is applicable to general situations, but it does not rule out the case of very individual counterexamples. You can give a list of various indices of a bunch of excited states when writing the paper, but at least it should also be at the same time Consider the hole and electron distribution diagram.

### 3.2.3  Exciton Binding Energy

The electron is negatively charged, and the place where the electron leaves, namely the hole, is correspondingly positively charged, so there is a Coulomb attraction energy between the electron and the hole, which is also called exciton binding energy, which can be calculated according to the simple Coulomb formula, as shown below:

$$E_C = \iint \frac{\rho^{\text{hole}}(\mathbf{r}_1)\,\rho^{\text{ele}}(\mathbf{r}_2)}{|\mathbf{r}_1 - \mathbf{r}_2|}\,\mathbf{dr}_1\mathbf{dr}_2 \tag{3.5}$$

However, taking the density corresponding to the occupied and unoccupied NTO with the largest eigenvalue as the hole and electron in the above formula is obviously not as ideal as the definition of hole and electron in Multiwfn, because in many cases the pair of NTO transition pairs that contribute the most The contribution of electronic excitation is often far less than 100%, and the deviation is so large that it cannot be ignored at all. At this time, the exciton binding energy calculated in this way is not accurate.

When Multiwfn [19] calculates the exciton binding energy according to the above formula, it currently uses a uniformly distributed grid point method. Even if the number of grid points is not particularly large, it is still very time-consuming. It is recommended to use a better performance server for calculation. Moreover, the time is proportional to the second power of the number of grid points, and inversely proportional to the sixth power of the grid point spacing. If the grid point spacing is too large, the quality of the grid points will be rougher. Although the time-consuming is low, the calculation accuracy is poor; the grid point spacing is too small and the accuracy is good, but the time-consuming will be high. Therefore, the grid point spacing should be selected appropriately, and the convergence test of the calculation result with the grid point spacing can be done at an appropriate time to see when the result basically does not change significantly with the increase in the number of grid points, then the number of grid points at this time is sufficient. In addition, there is another definition of exciton binding energy, which is the vertical ionization energy (VIP) minus the vertical electron affinity (VEA) and then the optical gap (the lowest electron excitation energy). This definition and the above definition are quite different in theory and actual values, and there is no significant connection.

### 3.2.4  Ghost-Hunter Index

The Ghost-hunter index itself does not belong to the category of hole-electron analysis, but since the aforementioned D index is used, and Multiwfn will also output this index during the hole-electron analysis process, I will mention it here. Pure functionals and hybrid functionals with low HF components are likely to have ghost excited states for large conjugated systems. This is due to the false charge transfer excited states caused by the self-interaction error (SIE) of the functional. The excitation energy of these excited states is very low, and the vibrator intensity is close to 0. Their existence will waste calculations and hinder correct judgment and discussion of excited states of interest (for example, beginners may mistake the ghost state for the S1 state to discuss fluorescence Emission characteristics). Generally, in order to avoid ghost states, high HF component functionals such as M06-2X, or long-range correction functionals such as wB97XD, or long-range functionals with high HF components such as CAM-B3LYP can be used to avoid them. In order to judge whether the ghost state may exist in the current TDDFT calculation. The definition of calculating the ghost-hunter index in the hole-electron analysis module of Multiwfn is:

$$M_{AC} = \sum_{i \to a} \frac{\left(\omega_i^a\right)^2 \left(-\epsilon_i - \epsilon_a\right)}{\sum_{i \to a} \left(\omega_i^a\right)^2} - \frac{1}{D} \tag{3.6}$$

In this formula, $\epsilon$ is the molecular orbital energy, i and a correspond to occupied and non-occupied orbitals, and D is the distance between the center of mass of the hole and the electron mentioned above. The ghost-hunter index is represented by the $M_{AC}$ symbol, and its corresponding full name is Mulliken averaged configuration. The ghost-hunter index is the lower limit of the excitation energy of the charge transfer excitation. If the excitation energy of a state calculated by TDDFT is lower than the corresponding ghost-hunter index, the state is considered a ghost state, otherwise it is not a ghost state. This judgment method has physical meaning and rationality, but according to the author's actual test, it is found that the standard for judging ghost state is often too strict. Sometimes the calculated excitation energy is lower than the ghost-hunter index, but based on basic theoretical knowledge, it can be judged that this is actually not a ghost state. Therefore, the ghost-hunter index is for reference only. Don't use it blindly or be scared by it.

Note that the ghost-hunter index is calculated based on the expensive D index calculated by the difference between the relaxed excited state density and the ground state density. However, in the hole-electron analysis module of Multiwfn, the ghost-hunter index calculated by D based on the distribution of holes and electrons is used, so there is a quantitative difference from the original method. This implementation in Multiwfn is absolutely reasonable and can be used with confidence. If you want to calculate the ghost-hunter index based on the difference between the relaxed excited state density and the ground state density difference, you should use Multiwfn's density difference-based charge transfer analysis function to calculate D.

# Chapter 4
# Vibration Spectrum Calculation and Analysis

## 4.1 IR Spectra

In chemistry, infrared spectroscopy is often used to analyze the composition and changes of solutions, because some molecular groups have infrared characteristic fingerprints. The problem is that the solvent and solute peaks are often superimposed on top of each other, making analysis very difficult. Therefore, we can simulate the infrared spectrum of the solvent with the help of molecular dynamics simulation to help analyze the infrared spectrum of the entire solution.

To calculate the infrared spectrum of a substance, the easiest way is to use quantum chemistry to calculate a single molecule in the gas phase. That is, Gauss is usually used to run an optimization and frequency (opt + freq) calculation. This calculation is based on the harmonic approximation (spring approximation) to obtain the second derivative of the energy to the atomic displacement to obtain the frequency, and calculate each vibration mode according to the relationship between the absorption intensity and the square of the corresponding transition dipole moment The relative strength. This calculation has two drawbacks. One is that the simple harmonic approximation has large errors for systems with strong non-harmonic properties, such as water molecular clusters and flexible molecules with rotatable groups; second, there is no intermolecular interaction in gas phase single molecules. There are intermolecular effects in the solution, which can also cause large errors.

How to do it? Using condensed matter molecular dynamics to simulate the solution system, and then Fourier decomposition of the autocorrelation function of the polar evolution of the system, the infrared spectrum containing both the intermolecular interaction effect and the harmonic approximation can be obtained. So what is its principle? This starts with the absorption of infrared photons by matter.

Take a beam of infrared light with a specific frequency to irradiate the sample, then the infrared photon has a specific energy $E = h\nu$, $\nu$ is the frequency. If this energy is equal to the difference between the vibrational dynamics of a molecule in the sample, the molecule will absorb it and transition To the vibrational excited state.

© Tsinghua University Press 2023
M. Sun and X. Mu, *Computational Simulation in Nanophotonics and Spectroscopy*,
Nanoscience and Nanotechnology,
https://doi.org/10.1007/978-981-99-4732-4_4

Most of the excitations are the transition from the ground state to the first excited state, and the first excitation energy is exactly equal to the ground state vibration energy, that is, the ground state frequency = absorption frequency. Therefore, in MD simulation, only the ground state of the solution needs to be simulated and the excited state is not required.

In addition to frequency, we also need to know the absorption intensity. What determines the absorption intensity? The first is of course the concentration. The more molecules there are, the stronger the absorption. The second is the transition dipole moment, that is, the dipole moment change caused by vibration. Only the vibration mode that causes the dipole moment to change can absorb infrared photons. Photons are composed of mutually perpendicular electric and magnetic fields. When the direction of the electric field is consistent with the direction of the transition dipole moment of a certain vibration mode of the molecule, effective absorption is possible. Therefore, the vibration that does not change the molecular dipole moment will not produce Absorbed. In addition, it is also related to the refractive index of the solution system, and this refractive index changes with frequency.

Therefore, if we want to detect the vibrational frequencies of the molecules in the solution that cause infrared absorption, we need to find an observable or calculable property. The microscopic quantity of this property can be related to the infrared absorption. This property is the box (the simulated ideal System) total dipole moment. Here requires a little imagination: the molecules and atoms in your box are constantly squirming randomly, and each squirming causes a change in the dipole moment of the related atomic group (imagine the expansion and contraction of a $C=O$ bond), all groups The change of dipole moment formed by creep constitutes the total dipole change. We record the value of the total dipole moment at each moment, and then convert the dipole moment signal into a frequency-intensity signal to obtain the infrared spectrum. How to convert? There is a mathematical tool called Fourier Transform, which specifically transforms time domain signals into frequency domain signals.

The experimental spectrum of a substance is a continuous absorption curve, including a large number of absorption peaks, showing the difference in the absorbance of light of different frequencies. It is generally measured by the molar absorption coefficient (epsilon, $\epsilon$), which means that the solution concentration is 1 mol/L, the absorbance when the thickness of the liquid layer is 1 cm, the unit is L/mol/cm.

The theoretical spectrum calculation gives discrete transition data. For example, when calculating the electronic excitation, the program will give the transition energy from the ground state to each electronic excited state and the oscillator strength. For example, in the figure below, the abscissa position of the black vertical lines is the electron excitation energy, and the height of the vertical line is the vibrator intensity (corresponding to the right coordinate axis).Obviously, the discrete transition data given by theoretical calculations are completely different from the continuous absorption curve given by experiments. If you want to correspond to the experimental spectrum to predict the actual spectrum or explain the experimental spectrum, you need to perform the theoretical transition data "Broadening" becomes a peak shape. First widen each transition. For example, the vertical line at 160 nm in the figure corresponds to the transition of $S_0 \rightarrow S_2$. After widening, it will be a red curve. After $S_0 \rightarrow S_{11}$ is widened, it will be a pink curve (because there are many transition methods, in order to avoid too chaotic, The figure only shows the curve after broadening the transition with the vibrator intensity greater than 0.01).

Each peak of the absorption curve often corresponds to a transition of greater intensity. For example, there is a peak around 113 nm in the figure, which obviously corresponds to the transition of $S_0 \rightarrow S_{13}$ (the oscillator intensity is 0.129). However, the position of the peak and the energy of the transition with greater intensity do not always correspond, whether it is the actual spectrum or the spectrum theoretically simulated by the above method, because all transitions contribute to the absorption curve in the nearby range. For example, the transition oscillator intensity of $S_0 \rightarrow S_3$ in the above figure is not very small, it is 0.036, but the black curve has no peak at the corresponding position (146.3 nm). The reason is easy to understand from the figure above. This is because there is a transition $S_0 \rightarrow S_5$ with a higher oscillator intensity near $S_0 \rightarrow S_3$ (the oscillator intensity is 0.107), and its contribution to the spectrum (shown by the cyan curve) is compared to The contribution of $S_0 \rightarrow S_3$ (shown in the blue curve) is much larger, which causes $S_0 \rightarrow S_3$ to have no corresponding peak, but is submerged in the absorption peak with the maximum at 138.2 nm. This example also illustrates the usefulness of examining the contribution of each transition to the spectrum. Assuming that we don't draw the contributions of these transitions with high vibrator intensity separately, the internal structure of the spectrum is not easy to figure out, and even leads to peaks. The essence of wrong identification. This function of drawing independent contributions of each transition is not available in non-professional programs such as GaussView, and can be realized in Multiwfn.

The simple example above is about ultraviolet spectroscopy. The process of generating spectra is the same for infrared, ECD (electronic circular dichroism), and VCD (vibrating circular dichroism). Theoretical calculations will give the energy and intensity values of each transition, and each of them must be widened as a curve and added. The intensity given by different types of spectrum calculations has different names. Infrared is called infrared intensity, UV–Vis is called vibrator intensity, and VCD/ECD calculation is called rotor intensity (positive and negative, corresponding to the left and right circularly polarized light). In the following we collectively refer to the "strength" of the transition.

PS: Let me talk about the unit by the way. For infrared, Raman and VCD spectra, the commonly used units are $cm^{-1}$; for UV–Vis and ECD, the commonly used units are $1000 \, cm^{-1}$, eV and nm. $1 \, eV = 8.0655 * 1000 \, cm^{-1}$. The reciprocal of the energy in eV multiplied by 1240.7011 is the energy in nm, and the reciprocal of the energy in nm multiplied by 1240.7011 is the energy in eV. Therefore, energy units such as nm and eV and $cm^{-1}$ are not linear The conversion relationship. The vibrator intensity is dimensionless. The infrared intensity unit is usually km/mol (kilometers per mole), and sometimes $1 \, esu^2 * cm^2 = 2.5066 \, km/mol$ is used as the unit. Gaussian output ECD rotor strength unit is cgs ($10^{-40}$ erg-esu-cm/Gauss), VCD rotor strength unit is $10^{-44} esu^2 \, cm^2$.

Given the energy and intensity of a transition, how can it be broadened into a peak-shaped curve as shown in the previous figure? This requires two conditions, one is the broadening function, which defines the functional form of the curve; the other is the full width at half maximum (FWHM), which determines the width of the broadened peak at half the height. HWHM is also used in many places, which means half of the width at half the height of the peak, HWHM=FWHM/2. Obviously, the larger the FWHM, the wider the peak looks, and the smaller the FWHM, the narrower the peak looks. If FWHM is an infinitely small value, then this is not a peak, but a vertical line, which is equivalent to No widening. The figure below shows the absorption curve when FWHM is different. There are three commonly used widening functions, Gaussian function, Lorentz function and Pseudo-Voigt function

$$
\begin{aligned}
&\text{Gaussian function} \quad G(\omega) = \frac{1}{c\sqrt{2\pi}} e^{-\frac{(\omega-\omega)^2}{2c^2}} \text{ where } c = \frac{\text{FWHM}}{2\sqrt{2\ln 2}} \\
&\text{Lorentzian function} \quad L(\omega) = \frac{\text{FWHM}}{2\pi} \frac{1}{(\omega-\omega_i)^2 + 0.25 \times \text{FWHM}^2} \\
&\text{Pseudo-Voigt function} \quad P(\omega) = w_{\text{gauss}} G(\omega) + \left(1 - w_{\text{gauss}}\right) L(\omega)
\end{aligned}
\tag{4.1}
$$

Generally, electronic spectroscopy (UV–Vis, ECD) uses Gaussian function to broaden, vibration spectrum (infrared, Raman, VCD) uses Lorentz function to broaden, Lorentz function decays more slowly than Gaussian function. The Pseudo-Voigt function is a linear combination of the Gaussian function and the Lorentz function, and the combination coefficient is adjustable. In the above formula, $\omega$ is the abscissa of the spectrum. After the transition energy $\omega$ and FWHM are given, there is an expression of the absorption curve corresponding to this transition immediately. The peak is highest at the position of $\omega = \omega_i$, and the function value gradually attenuates as $\omega$ deviates from $\omega_i$. The functions given above are all normalized, which means that the integral value of the function is 1. Of course, the intensity of the transition should also be taken into consideration when widening.

We have the mathematical form of the broadening function, and we also know the proportional relationship between the peak area and the transition intensity. How can the broadened absorption curve quantitatively correspond to the experimental spectrum? For the infrared spectrum, by comparing the infrared intensity unit and the molar absorption coefficient unit, it can be inferred that if $cm^{-1}$ is the abscissa unit and L/mol/cm is the ordinate unit, then If the infrared intensity of a transition is p, then the area of the curve expanded should be 100p, that is to say, the expansion function can be multiplied by 100p. For other types of spectra, there is no such formal relationship. We can only simulate the shape of the spectrum. Its value differs from the actual spectrum by a coefficient factor. This factor can only be obtained by comparing a large number of experimental spectra with simulated spectra. Fortunately, someone has done this for UV–Vis, and the conclusion is: if $1000 \, cm^{-1}$ units are used for the horizontal axis of the spectrum, and L/mol/cm units are used for the vertical axis of the spectrum, then 1 unit of vibrator. If eV is taken as the horizontal axis unit, the intensity of 1 unit of the vibrator should be expanded out of the curve with an area of 28700. With this relationship, the theoretically simulated UV–Vis spectrum is quantitatively comparable with the experimental spectrum. However, there is no such relationship for Raman, VCD, and ECD. Therefore, the transition of p intensity can only be simply expanded to the peak of integration area p, and the scale coefficient must be adjusted to make the simulated spectrum coincide with the experimental spectrum.

The transitions calculated by quantum chemistry theory are all discrete. Why is the measured spectrum continuous instead of absorption only when the incident frequency is appropriate and the excitation energy is exactly the same, so that discrete spectral lines can be observed? This problem is introduced in some molecular spectroscopy books. There are many reasons for the limited width of the spectrum, such as (1) the broadening caused by the uncertainty principle, that is, the excited state lifetime is limited, so the excited state energy is uncertain (2) Broadening caused by the Doppler effect caused by the movement of molecules (3) Broadening caused by energy level shifts caused by intermolecular collisions (4) High radiation intensity leads to saturation caused by depletion of the population of low energy states Broadened (5) Flexible molecules have a large number of accessible conformations.

The theoretical simulation spectrum and the experimental spectrum often have a certain overall deviation. In order to match as much as possible, we often need some adjustments. One is to scale the height of the spectrum, that is, multiply it by the scale factor, so that the peak height of the simulated spectrum corresponds well to the experimental spectrum. Generally, it is more difficult to calculate the intensity of the quasi-spectrum than the position of the quasi-peak. It is good to be able to qualitatively match it. Moreover, as mentioned above, there is no theoretical correspondence between the simulated spectrum and the experimental spectrum, so such a high-level scale It is completely reasonable and necessary. In addition, the abscissa of the simulated spectrum is also scaled or added or subtracted as a whole to eliminate the systematic deviation of the calculation of the transition energy. For example, the excitation energy calculated by CIS is usually too high. In some studies, it will be corrected by multiplying by 0.72. Vibration spectroscopy is known to also need to be multiplied by a frequency correction factor to solve the systematic error of the calculation method and equivalently consider the non-resonance effect. For more information, please refer to "Talking about Resonance Frequency Correction Factor" (http://sobereva.com/221). In addition, sometimes it is necessary to adjust the FWHM and the stretching function to make the result closer to the experimental spectrum. This kind of adjustment is not considered fraud, because the problems involved are difficult to overcome or impossible to overcome. This is just to adopt some techniques in order to better analyze and interpret the experimental spectrum.

## 4.2 Raman Spectra

### 4.2.1 Spontaneous and Resonance Raman Spectra

Raman spectroscopy is a spectroscopic technique used to study the vibration mode, rotation mode and other low frequency modes of a system of crystal lattices and molecules. Raman scattering is an inelastic scattering. The laser range usually used for excitation is visible light, near-infrared light or near the near-ultraviolet light range. The laser interacts with the phonons of the system, causing the final photon energy to increase or decrease, and the phonon mode can be known from the changes in these energy. This is similar to the basic principle of infrared absorption spectroscopy, but the data results obtained by the two are complementary. Usually, a sample is irradiated by a laser beam, and the irradiated spot is focused by a lens and split by a spectrometer. When the wavelength is close to the wavelength of the laser, it is elastic Rayleigh scattering. Spontaneous Raman scattering is very weak, and it is difficult to separate the Rayleigh scattering, which has a high intensity relative to the Raman scattering, so that the obtained result is a weak spectrum, which makes the measurement difficult. Historically, Raman spectrometers have used multiple gratings to achieve a high degree of light splitting and remove the laser light, while obtaining small differences in energy. In the past, the photomultiplier tube was chosen as the detector of Raman scattering signals, and it took a long time to get the result. Today's technology, notch filters can effectively remove laser light, and the advancement of spectrometers or Fourier transform spectrometers and charge-coupled device (CCD) detectors. In scientific research, Raman spectroscopy is used to study materials. Features are becoming more and more extensive. There are many kinds of Raman spectroscopy, such as surface enhanced Raman effect, tip enhanced Raman effect, polarized Raman spectroscopy and so on.When light hits the molecule and interacts with the electron cloud and molecular bonds in the molecule, the Raman effect occurs. For spontaneous Raman effects, photons excite molecules from the ground state to a virtual energy state. When the excited molecule emits a photon, it returns to a rotation or vibration state different from the ground state. The energy difference between the ground state and the new state makes the frequency of the released photons different from the wavelength of the excitation light. If the molecules in the final vibration state have higher energy than the initial state, the frequency of the photons excited will be lower to ensure the total energy balance of the system. This frequency change is called Stokes shift. If the molecule in the final vibration state has lower energy than the initial state, the photon frequency excited will be higher. This frequency change is called Anti-Stokes shift. Raman scattering is the transfer of energy through the interaction between photons and molecules, which is an example of inelastic scattering. Regarding the coordination of vibration, the change of molecular polarization potential or the change of electron cloud is the inevitable result of molecular Raman effect. The amount of change in the polarization rate will determine the intensity of Raman scattering. The change of the mode frequency is determined by the rotation and vibration state of the sample.

Raman spectroscopy is widely used in the field of chemistry because chemical bonds and symmetrical molecules have their special vibrational spectral information, so they provide important features for molecular identification. For example, the vibration frequencies of SiO, $Si_2O_2$, and $Si_3O_3$ can be identified, and are listed as the basis of infrared spectroscopy and Raman spectroscopy coordination analysis. The special (wave number) range of organic molecules is 500–2000/cm. On the other hand, spectroscopy coordination analysis technology has also been applied to the study of chemical bonding, for example, adding enzymes to the substrate. Raman gas detectors have many practical applications. For example, in medicine, the real time for anesthetics to work and the real time for mixed breathing gas during surgery. Spontaneous Raman spectroscopy is often used in solid-state physics, such as raw material characteristics, temperature measurement and crystallographic orientation of samples. For example, the special phonon patterns of a group of solid materials allow the experimenter to quickly identify single crystals. In addition, Raman spectroscopy can monitor solid-state low-frequency excitations, such as plasma, magnon and superconducting gas excitation. The Raman signal provides information on the ratio of the intensity of Stokes (low frequency conversion) to the intensity of anti-Stokes (high frequency) in the phonon mode. Raman scattering is generated by anisotropic crystals and provides information to determine crystal orientation. The polarization of Raman light depends on the polarization of the crystal and laser. If the crystal structure (especially, the point group of the crystal structure) is known, it can be used to find the direction of the crystal.

## 4.3 Calculation of Vibration-Resolved Electronic Spectra

### 4.3.1 Principles

On the surface, both photoelectron spectroscopy and UV–Vis are only spectra of changes between electronic states (this article only discusses UV–Vis absorption spectra), and the absorption peak comes from the transition between electronic states. But in fact, each electronic state also corresponds to many vibration modes. For example, in the spectrum of the transition from the electronic state of A to the electronic state of B, as long as there is light of the corresponding frequency, the electrons will actually transition from the vibrational ground state of A to the various vibrational states of B, and their transition energies are different. Therefore, a peak of electronic state transition, if the spectral resolution is increased to obtain a fine structure, it will be seen that it is composed of many vibration-related peaks. This is called vibrationally resolved electronic spectra.

The system will be in a vibrational ground state at $0\,$K. At a limited temperature, the vibrational excited state of A will also have a certain distribution, so it can also transition from the vibrational excited state of A to the various vibrational states of B. According to the Boltzmann distribution, the distribution ratio of each vibrational state of A is obtained, and the spectrum of each vibrational state of A transitions to B is superimposed by weight, which is the vibration-resolved electronic spectrum observed at the actual temperature. Therefore, the temperature dependence of the vibration-resolved electronic spectrum can be calculated theoretically.

Theoretical calculation of vibration-resolved electronic spectra needs to consider the transition between the "electron + nucleus" wave function $\Psi$ of electronic ground state $\nu =\mid 0\rangle$ to various excited states $\nu =\mid 0\rangle$, where v represents the vibration quantum number, and 0 corresponds to the vibration ground state. Perform vibration analysis in the ground state and excited state tasks to obtain the vibration energy levels in the two electronic states, and calculate the difference to obtain the transition energy between the various states involved in the vibration-resolved electronic spectrum. But it's useless to know this. In order to make a graph, what we need is the oscillator strength of each such transition. This requires knowing the transition dipole moment between each $\Psi$. The oscillator strength is proportional to the transition dipole. The square of the moment. Under the BO approximation, the transition dipole moment $\langle \Psi' \mid \mu \mid \Psi'' \rangle$ can be separated into the electronic wave function $\Phi$ and the nuclear wave function $\Psi$ part: $\langle \Psi' \mid \mu \mid \Psi'' \rangle = \langle \Psi' \mid \mu_e \mid \Psi'' \rangle$>, where $\mu_e$ is the electronic transition dipole moment $\langle \Phi' \mid \mu \mid \Phi'' \rangle$.

$\mu_e$ is obviously dependent on the nuclear coordinates, and can be expanded by Taylor relative to the excited state equilibrium structure. Processing of it leads to three methods of $\langle \Psi' \mid \mu \mid \Psi'' \rangle$ calculation:

(1) FC (Franck–Condon) approximation: $\mu_e$ only takes the first term expanded by Taylor, so $\mu_e$ a constant, that is, the electronic transition dipole moment of the excited state equilibrium structure. For the study of bright states, this assumption is usually sufficient to give reasonable results.

(2) HT (Herzberg–Teller) method: only take the second term expanded by Taylor, so $\mu_e$e is a function of nuclear coordinates. This method is usually not used alone, because the result is definitely not in line with the actual situation. After all, the first item Taylor developed is the most important. The use of HT alone is only to discuss the influence of the Herzberg–Teller effect on the vibration-resolved electronic spectrum. However, sometimes the two states are strictly forbidden from the electronic transition dipole moment due to the symmetry of the electronic states. However, if nuclear vibration is also taken into account by the HT method, the two states The dipole moment of the interm transition is no longer zero, which makes the transition have a certain (but small) chance to occur. So if you want to study very weak transitions, especially dark states, HT must be considered.

(3) FCHT method: both FC and HT are counted, and both the first term (constant term) and the second term (first-order correction) of Taylor's expansion are considered. This result is obviously wider than the FC approximation.

PS: Don't confuse the FC principle, FC factor (or FC integral), and FC approximation. Although they are related, the specific issues are different. The FC principle means that the electronic transition process is very short, and the nuclear coordinates are too late to change. The FC factor refers to the square of the overlap integral between the ground state vibration wave function and the excited state vibration wave function. The FC approximation is a simplified processing method for calculating the transition dipole moment used in the vibration-resolved electronic spectrum, assuming that $\mu_e$ is a constant and is not affected by nuclear coordinates.

There are 3N-6 normal vibration modes in a non-linear polyatomic molecule, and they can be regarded as independent under the harmonic oscillator model, that is, the vibration wave function of the system can be written as the product of the wave function of each vibration mode. If the vibration quantum number v of each vibration mode is 0, then the system is in the vibration ground state. If one vibration mode v > 0, or multiple vibration modes v > 0 at the same time, then the system is in a vibration excited state. Because v has no upper limit in the resonance model, and there can be various combinations between the excited states of these vibration modes, the number of vibration states of the electronic excited state is very large, and the vibrator intensities from the ground state to all these states are considered It is impossible, the calculation is too large. Fortunately, if the FC factor between the two states is small, then the transition dipole moment between them will also be small, so it has no effect on the spectrum and can be ignored. In addition, you can also specify the study range of the spectrum. If the energy difference between the two states exceeds this range, there is no need to calculate the oscillator intensity between them. FCClasses is a systematic pre-screening method that estimates and classifies transitions in advance, and only counts transitions between states that are important to the actual spectrum. The screening threshold corresponds to the amount of calculation. Generally, it is enough to adjust the number of points to be calculated, which belongs to the black box method.

## 4.3.2  Calculation Methods

Gaussian09 can use Freq keywords to calculate vibration-resolved electronic spectra. In fact, it embeds the FCClasses program (http://village.pi.iccom.cnr.it/Software) that specifically analyzes this problem. Either CIS or TDDFT can be used to calculate the excited state. Although in principle the non-resonance model can be used to calculate the vibration-resolved electronic spectrum, but currently only supports the calculation of the vibration-resolved electronic spectrum at the resonant frequency, because the non-resonant effect needs to consider the third or higher order derivative, Gaussian's excited state calculation cannot Up to this point (CIS can do second-order analytic derivatives, TD can only do first-order analytic derivatives).

The following uses anisole as an example to illustrate how to calculate the vibration-resolved electronic spectrum of $S_0 \rightarrow S_1$.

Use the following input file to optimize the $S_1$ excited state and perform vibration analysis. The saveNM keyword causes the excited state vibration information to be stored in chk, and generates the electronic transition dipole moment of the excited state equilibrium structure required by the FC method discussed above when calculating $\mu_e$. If you want to study which electronic excited state.

```
%chk=C:\gtest\anisole_exc.chk
# cis(root=1)/6-31G* opt freq=saveNM

anisole S1

0 1
C                  2.28445000      0.32691600      0.00003800
C                  1.34343300      1.38142800     -0.00010400
C                 -0.03741900      1.09055200     -0.00003400
C                 -0.44320000     -0.26595700     -0.00005800
C                  0.49751700     -1.32326400     -0.00002000
C                  1.87678500     -1.02037200      0.00017600
```

```
H                  3.33003700      0.56133100     -0.00002300
H                  1.67739200      2.39758400     -0.00018600
H                 -0.75550500      1.88022800      0.00018000
H                  0.12880400     -2.32514600     -0.00027400
H                  2.60234200     -1.80568700      0.00035800
O                 -1.73605800     -0.64937600     -0.00027400
C                 -2.82140400      0.30027500      0.00020300
H                 -3.71931200     -0.29382400      0.00075700
H                 -2.78766600      0.92165900      0.88463800
H                 -2.78859800      0.92139000     -0.88445500
```

Then use the following input file to optimize the ground state and make a vibration analysis of the ground state. The freq=FC task will use the FC method to calculate the transition energy and transition dipole moment from the vibration ground state of $S_0$ to the vibration states of $S_1$ based on the vibration information of the current $S_0$ state and the vibration information in the chk file of the $S_1$ state just now. And converted into vibrator intensity.

```
%chk=C:\gtest\anisole.chk
# hf/6-31G* opt freq=FC nosymm

anisole S0

0 1
C                 -2.27936800      0.32609200      0.00008100
C                 -1.34379900      1.33903700     -0.00003400
C                  0.01378100      1.05103000     -0.00016100
C                  0.43663600     -0.26434700     -0.00020100
C                 -0.50250800     -1.28814500     -0.00003400
C                 -1.84702200     -0.99409200      0.00014300
H                 -3.32624700      0.55310300      0.00026000
H                 -1.66092300      2.36324200      0.00002000
H                  0.71962800      1.85453200     -0.00023700
H                 -0.14729200     -2.29705700     -0.00003400
H                 -2.56259500     -1.79229300      0.00028200
O                  1.75151800     -0.65529700     -0.00024100
C                  2.80692300      0.31779600      0.00026800
H                  3.72320000     -0.24943200      0.00040700
H                  2.76733600      0.94326200     -0.88309400
H                  2.76687800      0.94280000      0.88395800

C:\gtest\anisole_exc.chk
```

In the FC calculation process, the 0–0 transition (S0 vibration ground state -> S1 vibration ground state) energy will be output before outputting various transition modes

```
Energy of the 0-0 transition: 47860.6879 cm^(-1)
```

In fact, this value can be obtained by subtracting the energy of the excited state and the ground state including ZPE. You will see in the output of the excited state calculation

```
Sum of electronic and zero-point Energies= -344.222064
```

You will see in the output of the ground state calculation

```
Sum of electronic and zero-point Energies= -344.440134
```

Therefore $(-344.222064 + 344.440134) * 219474.6363 = 47860$ cm$^{-1}$.

In the output such as

```
Initial State: <0|
Final State: |15^2>
DeltaE = 1683.4671 | TDMI**2 = 0.1337E-01, Intensity = 0.4341E-01
```

It refers to the transition from the vibrational ground state of S0 to the vibrational excited state of S1. This vibration excited state corresponds to the situation where the No. 15 vibration mode is at v=2. DeltaE is the transition energy, not an absolute value, but a value relative to the 0–0 transition energy. TDMI is the square of the mode of the transition dipole moment.

Another example

```
Initial State: <0|
Final State: |26^1;17^1>
```

The vibrational excited state of S1 here is a combination of the situation where the vibration mode No. 26 is at v=1 and the vibration mode No. 17 is also at v=1.

Finally, Gaussian will output the spectrum simulated by Gaussian function broadening

```
+------------------+
| Final Spectrum |
+------------------+

Axis X = Energy (in cm^-1)
Axis Y = Intensity

-------------------------------------------------- ----------
...slightly
47476.6879 0.247451D-02
47484.6879 0.311828D-02
47492.6879 0.391045D-02
47500.6879 0.488005D-02
47508.6879 0.606049D-02
47516.6879 0.748991D-02
47524.6879 0.921152D-02
...slightly
```

Just use origin to directly plot the two columns of data, which is the vibration-resolved electronic spectrum, as shown by the black line in the figure below. The highest peak in the figure corresponds to the 0–0 transition.

### 4.3.3  Additional Parameters

If freq=(FC,ReadFCHT) is used in the calculation of the ground state, the additional parameters of the control calculation will be read. See the end of the Freq keyword in the Gaussian manual. The more important ones are MaxOvr: The maximum number of vibrational quanta in the excited state is considered. The default is 20. For electronic transitions with large geometric changes, vibrational excited states with higher quantum numbers are involved, and the spectral range will be relatively wide. If this value is not set at this time, the spectrum will be incomplete. MaxInt: The number of points to be calculated for each type of transition, in millions, the default is 100, which is $10^8$. The time-consuming calculation of FC/HT/FCHT does not lie in the system size, basis set, and theoretical method, but only in the number of points to be calculated. Therefore, the larger this value is set, the more accurate the obtained spectrum, but the greater the amount of calculation. SpecHwHm: The half-height and half-width used when the calculated result is expanded into the spectrum, the default is 135 ($cm^{-1}$). PrtInt: The default is 0.01, that is, if the vibrator intensity of the transition mode is greater than 1% of the vibrator intensity of the 0–0 transition, the transition mode is output in the output file. DoTemp: Whether to consider temperature when calculating spectrum, write DoTemp to consider temperature. The default is 0 K, so the initial state only considers the vibrational ground state of the electronic ground state. Note that although this keyword exists, it does not take effect in Gaussian. NORELI00 SPECMIN=37900 SPECMAX=42000: This means that only the part of the transition energy from 37900 to 42000 $cm^{-1}$ is calculated. SPECRES: The default is 8($cm^{-1}$). As you can see in the output of the above example, 47476.6879, 47484.6879, 47492.6879... are all separated by 8. The smaller the value, the higher the spectral accuracy, but the more time-consuming the calculation. PrtMat: If set to 1, the Duschinsky matrix J will be output; if set to 2, the displacement vector K will be output. (Note: Duschinsky rotation or Duschinsky mixing effect is the electronic transition that causes linear mixing (rotation) of the ground state vibration mode to produce an excited state vibration mode, expressed as $Q'' = J Q' + K Q''$ And $Q'_i$ respectively represent the i-th normal vibration coordinate of the electronic excited state and the electronic ground state. If the vibration modes of the ground state and the excited state are exactly the same, then J is the identity matrix, which means there is no mixing.)

In addition, it is well known that there is a certain difference between the resonant frequency and the non-resonant frequency, and the frequency correction factor is usually used to correct the resonant frequency. The SclVec keyword can also be used to correct the resonant frequencies of the ground state and the excited state to approximate the vibration-resolved electronic spectrum of the non-resonant model. Specifically, it provides the non-resonant calculation of each vibration frequency of the electronic ground state (G09 supports the PT2 method to obtain the non-resonant frequency), and divides the result of the current resonance model calculation to obtain the correction factor corresponding to each vibration mode. Through the correction factor of each vibration mode of the ground state and mixing according to the Duschinsky matrix, the correction factor of each vibration mode of the excited state is obtained. Of course, it is obviously more accurate to directly calculate the non-resonant frequency of the excited state to obtain the non-resonant frequency, but the non-resonant calculation requires a higher order derivative, which is difficult to achieve for the excited state, so this method can only be used to indirectly correct the factor by the ground state frequency. The correction factor for the vibration frequency of the excited state is estimated.

## 4.4 Vibration Mode

In Gaussian, there is an option intmodes that is useful for analyzing vibration modes in the freq keyword. Unfortunately, its description in the manual is very obscure, causing many people to not notice it. Here is a brief introduction.

freq=intmodes can decompose each normal vibration mode into the contribution of each redundant internal coordinate. This can quantify the discussion of vibration modes and can also help identify the characteristics of vibration modes. Here we take ethanol under HF/6-31G* as an example. The animation below is the second vibration mode of ethanol.

After using freq=intmodes, after the thermodynamic data output, you can find the components of each redundant internal coordinate in this vibration mode.

```
         -----------------------------
! Normal Mode     2        !
         -----------------------------           --------------------------------
! Name  Definition           Value        Relative Weight (%)            !
-----------------------------------------------------------------------------
! D1      D(2,1,5,6)          -0.0977              3.5                    !
! D2      D(2,1,5,7)          -0.0977              3.5                    !
! D3      D(2,1,5,8)          -0.1078              3.9                    !
! D4      D(3,1,5,6)          -0.1015              3.7                    !
! D5      D(3,1,5,7)          -0.1015              3.7                    !
! D6      D(3,1,5,8)          -0.1116              4.0                    !
! D7      D(4,1,5,6)          -0.0977              3.5                    !
! D8      D(4,1,5,7)          -0.0977              3.5                    !
! D9      D(4,1,5,8)          -0.1078              3.9                    !
! D10     D(1,5,8,9)           0.6058             21.9                    !
! D11     D(6,5,8,9)           0.5882             21.3                    !
! D12     D(7,5,8,9)           0.5882             21.3                    !
         -----------------------------------------------------------------------
```

It can be seen that the three dihedral angles D(1,5,8,9), D(6,5,8,9) and D(7,5,8,9) account for a large component, accounting for a total of about 65%. Compared with the animation, this value is indeed more reasonable. These dihedral angles from D1 to D9 all correspond to methyl twist, accounting for about 33% in total. Therefore, when describing this vibration mode, you can write 33% $\tau$(C–C), 65% $\tau$(C–O), where the $\tau$ symbol is usually used to represent the torsion term.

By looking at the contribution of the redundant internal coordinates, you can easily imagine the vibration mode without watching the animation, such as the 19th mode of ethanol.

```
-------------------------------
! Normal Mode     19      !
-------------------------------                    -----------------------------
! Name  Definition           Value         Relative Weight (%)              !
-------------------------------------------------------------------------------
! R1    R(1,2)               -0.5164              20.6                       !
! R2    R(1,3)                0.8206              32.8                       !
! R3    R(1,4)               -0.5164              20.6                       !
! R5    R(5,6)                0.056                2.2                       !
! R6    R(5,7)                0.056                2.2                       !
! A2    A(2,1,4)              0.0474               1.9                       !
! A5    A(3,1,5)             -0.0474               1.9                       !
! D1    D(2,1,5,6)           -0.0358               1.4                       !
! D2    D(2,1,5,7)           -0.0508               2.0                       !
! D3    D(2,1,5,8)           -0.0433               1.7                       !
! D7    D(4,1,5,6)            0.0508               2.0                       !
! D8    D(4,1,5,7)            0.0358               1.4                       !
! D9    D(4,1,5,8)            0.0433               1.7                       !
-------------------------------------------------------------------------------
```

The three internal coordinates that account for the largest component are all C–H bond terms (74% in total), where the Value of R2 is positive, and R1 and R3 are negative, that is, the phase is opposite. Therefore, the main feature of this vibration mode is the asymmetric stretching vibration of the methyl group.

When freq=intmodes is used, redundant internal coordinates can be customized, which brings great flexibility to the study of the components of vibration modes. For example, in the animation of the second vibration mode of ethanol, it can be seen that the distance between atoms 7 and 9 changes greatly during the vibration process. Therefore, we can try to add a redundant internal coordinate between atoms 7 and 9 to understand its contribution to this vibration mode. The method is to write freq=(modredundant,intmodes), and then write B 7 9 in a blank line after the molecular coordinates. The output at this time is

```
-------------------------------
! Normal Mode      2      !
-------------------------------                    -----------------------------
! Name  Definition           Value         Relative Weight (%)              !
-------------------------------------------------------------------------------
! R8    R(7,9)                0.379               12.0                       !
! D1    D(2,1,5,6)           -0.0977               3.1                       !
! D2    D(2,1,5,7)           -0.0977               3.1                       !
! D3    D(2,1,5,8)           -0.1078               3.4                       !
! D4    D(3,1,5,6)           -0.1016               3.2                       !
! D5    D(3,1,5,7)           -0.1016               3.2                       !
! D6    D(3,1,5,8)           -0.1116               3.5                       !
! D7    D(4,1,5,6)           -0.0977               3.1                       !
! D8    D(4,1,5,7)           -0.0977               3.1                       !
! D9    D(4,1,5,8)           -0.1078               3.4                       !
! D10   D(1,5,8,9)            0.6057              19.3                       !
! D11   D(6,5,8,9)            0.5882              18.7                       !
! D12   D(7,5,8,9)            0.5882              18.7                       !
-------------------------------------------------------------------------------
```

# Chapter 5
# Calculation of Nonlinear Optical Properties

## 5.1 Two-Photon Absorption (TPA)

Two-photon absorption refers to the phenomenon that an atom or molecule absorbs two photons at the same time and transitions to a higher energy level. In this case, the energy difference between the energy levels is exactly equal to the total energy of the absorbed photon. Two-photon absorption requires two photons to react with the molecule at the same time, so the reaction probability is much smaller than the general single-photon absorption, and its probability is proportional to the square of the light intensity, so it belongs to the category of nonlinear optics. The discussion on two-photon absorption can be traced back to Gppert-Mayer's doctoral thesis in 1931 [20], but lasers had not yet been invented at that time, so it was difficult to achieve the light intensity required for two-photon absorption. The actual experiment was not realized until the 1960s. Multiphoton absorption may be accompanied by multiphoton emission and multiphoton effects such as conductance, photoelectricity, fluorescence, dissociation, and photochemical reactions. These phenomena in turn contribute to the study of multiphoton absorption processes. The multiphoton absorption process caused by laser irradiation has been successfully used for the separation of isotope sulfur, and it has important applications in the fields of spectroscopy, physical property research, isotope separation, and photochemistry. Two-photon absorption cross section and two-photon induced fluorescence are important characteristics of two-photon materials. At present, there are a variety of experimental methods for testing the two-photon absorption cross-section of materials, including nonlinear transmittance testing methods and transient two-photon induced fluorescence testing methods. The experimental method of transient fluorescence spectroscopy can be used to study two-photon induced fluorescence emission dynamics and excited state properties. Two-photon absorption has two important characteristics:

(1) Two-photon absorption is the process of long-wave absorption and short-wave emission. The excitation light has a high transmittance to the medium, which can

M. Sun and X. Mu, *Computational Simulation in Nanophotonics and Spectroscopy*,
Nanoscience and Nanotechnology,
https://doi.org/10.1007/978-981-99-4732-4_5

effectively reduce the dissipation and damage of the medium to the excitation light absorption;

(2) The absorption intensity is proportional to the square of the incident light intensity. In the case of tight focus, two-photon absorption only occurs in the space volume of the order of magnitude at the focal point! Based on the above characteristics, materials with large two-photon absorption cross-sections Many fields such as photon fluorescence microscopy and imaging, three-dimensional optical information storage, optical micromachining, frequency up-conversion lasing, optical limiting, and photobiology show good application prospects (see Fig. 5.1).

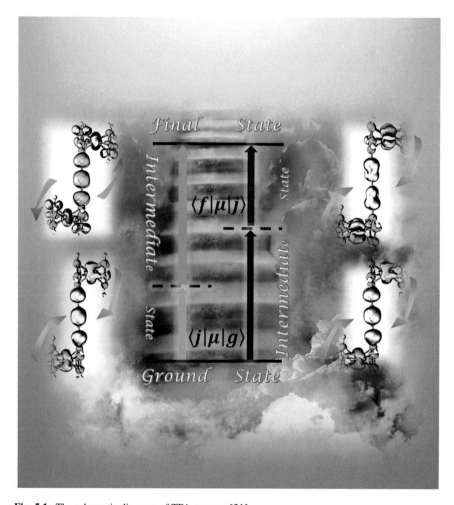

**Fig. 5.1**  The schematic diagram of TPA process [21]

Frequency up-conversion lasing refers to amplified spontaneous emission or cavity lasing generated under two-photon excitation. Compared with coherent frequency up-conversion such as harmonic generation and parametric mixing, up-conversion lasing has the following advantages:

(1) It does not require phase matching, and it can easily achieve wide-range tuning;
(2) It can be applied to waveguides and organic Optical fiber system! Compared with the traditional single-photon pumped lasing in organic materials, the advantage of the two-photon pumped lasing is that the pumping light wavelength is the wavelength of the common solid laser, and there is no need to double the frequency of the solid laser first. In recent years, two-photon induced fluorescence and upconversion lasing of organic materials have been actively studied.

In 1990, Cornell University Denk et al. proposed to apply the two-photon excitation phenomenon to confocal laser scanning microscopes [22], opening up a new field of two-photon fluorescence microscopy and imaging. Compared with single-photon confocal microscope, two-photon confocal microscope has many outstanding advantages:

(1) Two-photon confocal microscope can use infrared laser with relatively long wavelength and strong penetrating ability in biological tissues as excitation light source. Therefore, it can solve the problem of tomography imaging of deep-layer substances in biological tissues;
(2) Because the two-photon fluorescence wavelength is far away from the excitation light wavelength, the two-photon confocal microscope can achieve dark-field imaging;
(3) Two-photon fluorescence can avoid ordinary fluorescence imaging The problem of fluorescence bleaching and the phototoxicity to biological cells in the system;
(4) Two-photon transition has strong selective excitation, which is conducive to imaging research on some special substances in biological tissues;
(5) Two-photon confocal microscope It has higher horizontal and vertical resolution. In addition, because the two-photon absorption of the material is strongly related to the square of the excitation light intensity, under tightly focused conditions, the two-photon absorption is only localized in the space at the focal point of the objective lens.

Based on the above advantages, in addition to the use of Long-wavelength light to excite the sample can avoid the damage of ultraviolet light to the sample and the use of complex ultraviolet. Many limitations of optical components can extend the observation time of living biological samples. Two-photon confocal laser scanning microscope provides a unique and important method for studying amino acids, proteins and neurotransmitters. For example, Denk et al. use a wavelength of 630 Two-photon excitation with nm laser can obtain the fluorescence image of the chromosomes of pig kidney cells during the division phase with a resolution of 200 nm. Therefore, two-photon fluorescence microscopy has become a research for biomolecule detection and super-resolution tomography of active cells. An important tool. Conventional two-photon confocal microscopy and imaging usually use traditional fluorescent substances such as fluorescein as the two-photon fluorescent emitter. The two-photon absorption cross section of these fluorescent substances is small, and high-intensity laser excitation must be used in the experiment The measure of high concentration of fluorescent substance to obtain sufficiently strong up-conversion fluorescence, which increases the difficulty of observation of living biological samples. Therefore, seeking materials with high fluorescence quantum yield and large two-photon absorption cross section is a hot spot in this field. The deepening of these studies will greatly promote the development of two-photon confocal microscopy and imaging technology.

### 5.1.1   Calculation Method of TPA Cross-Section

The most common calculation method of TPA cross-section is the second response theory (QRT). This method is also included in most commercial programs. The more well-known program is Dalton. QRT theory has been tested for many years, and its calculation results are relatively close to experiments, which has certain application value. The following input files can be used to calculate the TPA cross-section through QRT in Dalton:

```
**DALTON INPUT
.RUN RESPONSE
*PCM
.SOLVNT
H2O
.NEQRSP
*PCMCAV
**WAVE FUNCTIONS
.DFT
B3LYP
**RESPONSE
*QUADRATIC
.TWO-PHOTON
.ROOTS
6
**END OF DALTON INPUT
```

The "RUN RESPONSE" is the keyword of response theory. The "SOLVNT" keyword can specify the type of solvent to be calculated. From "WAVE FUNCTIONS", you can specify the calculation method ("DFT" and "B3LYP"), response theory details ("QUADRATIC"), calculation purpose and related details ("ROOT").The corresponding TPA section in the output file is as follows:

```
*******************************************************************
************ FINAL RESULTS FROM TWO-PHOTON CALCULATION ************
*******************************************************************

The two-photon absorption strength for an average molecular
```

```
orientation is computed according to formulas given by
P.R. Monson and W.M. McClain in J. Chem. Phys. 53:29, 1970
and W.M. McClain in J. Chem. Phys. 55:2789, 1971.
The absorption depends on the light polarization.
A monochromatic light source is assumed.

All results are presented in atomic units, except the
excitation energy which is given in eV and two-photon cross
section which is given in GM. A FWHM of 0.1 eV is assumed.

Conversion factors:
1 a.u. = 1.896788 10^{-50} cm^4 s/photon
1 GM = 10^{-50} cm^4 s/photon
+-------------------------------+
| Two-photon transition tensor S |
+-------------------------------+

---------------------------------------------------------------------------
Sym  No  Energy        Sxx       Syy       Szz       Sxy       Sxz      Syz
---------------------------------------------------------------------------
 1   1    9.97        -0.1       3.2      -3.0      -2.2       0.8      1.2
 1   2    9.97         3.5      -2.5      -1.0       0.5       1.5      2.1
 1   3    9.98         3.2      -0.6      -2.5      -0.7      -1.6     -2.3
 1   4   11.16        -4.0      14.8     -10.6      -9.0       3.7      5.1
 1   5   11.16       -13.1       8.8       4.7      -1.3      -7.0    -10.3
 1   6   11.17        14.2      -1.5     -12.5      -3.9      -5.7     -8.1
 1   7   11.26        -0.2      -7.0       7.0      -9.8       1.7     -1.3
 1   8   11.26         0.2       1.1      -1.3       1.9      10.0     -6.8
......
 1  57   20.88        11.2       1.0     -12.3      -3.6      -6.4    -12.8
 1  58   20.88        -0.8      12.3     -11.5     -11.7       7.5      5.1
 1  59   20.89        -0.1      13.9     -13.8      14.4      -9.5      8.2
 1  60   20.89        -5.8      -1.6       7.3     -11.4     -17.4      7.3
---------------------------------------------------------------------------
```

In fact, the QRT theory has the problem that the calculation speed increases drastically as the number of atoms increases. In other words, the QRT theory is difficult to calculate for large systems. This problem can be solved by calculating the TPA section through SOS theory. Because for quantum chemistry programs, you only need to calculate enough excited states to easily calculate the TPA cross section. And this method based on SOS theory can analyze the intermediate state in the TPA transition process. This is because the calculation principle is defined as:

$$
\delta_{tp} = 8 \sum_{\substack{j\neq g \\ j\neq f}} \frac{|\langle f|\mu|j\rangle \langle j|\mu|g\rangle|^2}{\left(\omega_j - \frac{\omega_f}{2}\right)^2 + \Gamma_f^2} \left(1 + 2\cos^2\theta_j\right) + 8\frac{|\Delta\mu_{fg}|^2 |\langle f|\mu|g\rangle|^2}{\left(\frac{\omega_f}{2}\right)^2 + \Gamma_f^2}\left(1 + 2\cos^2\phi\right)
$$

$$(5.1)$$

In the formula, the first transition and the second transition can be separated, and the one-step transition and the two-step transition part in the TPA process can be separated and analyzed. Therefore, calculating the TPA process in this way can more conveniently analyze the details of the TPA transition process.

## 5.1.2   Application of TPA Calculation

Analysis of TPA transition characteristics is very important for designing new two-photon devices, molecules and materials. And we can explore the deep-seated physical mechanism of two-photon absorption regulation under special conditions. For example, the external electric field is very obvious in regulating the intensity of TPA between molecular dimers. In order to explore this regulatory mechanism, Mu et al. conducted a detailed analysis of this process. It can be seen from Fig. 5.2 that the external electric field can regulate different transition characteristics of the TPA transition process, including intermolecular and intramolecular charge transfer. It also has varying degrees of regulation on different TPA excited states of molecules, see Fig. 5.3.

The above example is to use the SOS method to explore the problem of external field regulation in the TPA process. This method can also explore the special charge transfer phenomenon in the TPA process from different angles. As shown in Fig. 5.4, it is a graph of the charge transfer density during the TPA transition of the porphyrin quarter thiophenes fullerene. From the figure, we can see significant super-exchange charge transfer and sequence charge transfer. This phenomenon comes from the start and end points between the two steps in the TDM diagram.

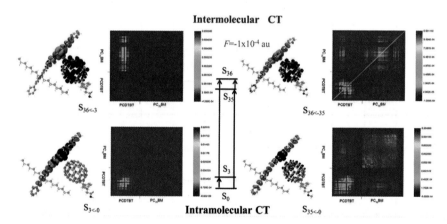

**Fig. 5.2** The electron-hole pair analysis isosurface diagram of the TPA process of donor-acceptor system controlled by external electric field [23]

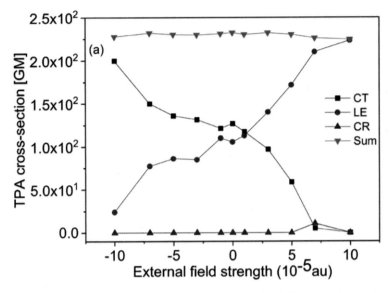

**Fig. 5.3** The external electric field controlled TPA charge transfer of different TPA states in donor-acceptor system [23]

## 5.2 Second Order Harmonic Wave Generate (SHG)

### 5.2.1 Sum-of-States (SOS)

The sum of complete states (SOS) is a common method to calculate the polarizability and hyperpolarizability. Multiwfn supports calculation of static/frequency-containing polarizability $\alpha$, first hyperpolarizability $\beta$, second hyperpolarizability $\gamma$ and even third hyperpolarizability $\theta$ through SOS. The energy and dipole moment of each state that needs to be used, as well as the transition dipole moment between each state, can be generated by Multiwfn based on Gaussian or ORCA's CIS, TDHF, and TDDFT calculation results. This article describes how to implement it.

Although Gaussian can directly use the TD (SOS) keyword to do SOS calculation based on the results of TDHF and TDDFT, it can only give the polarization rate, which is obviously too limited, and it does not support SOS calculation for commonly used CIS.

The specific principle of SOS will not be discussed here, but mainly introduce the actual calculation formula. The SOS calculation formulas for $\alpha$, $\beta$, $\gamma$, and $\theta$ can be found in J. Chem. Phys., 99, 3738 (1993). Note that the SOS formulas given in many documents and books are wrong. The formula for calculating the polarizability and the first hyperpolarizability is as follows [25]:

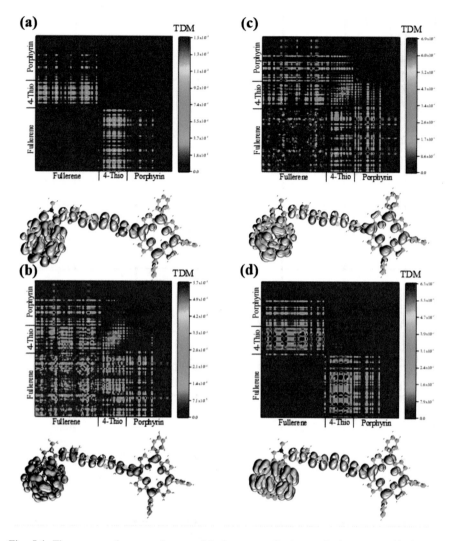

**Fig. 5.4** The super-exchange and sequential charge transfer in porphyrin quarter thiophenes fullerene [24]

$$\alpha_{AB}(-\omega; \omega) = \sum_{i \neq 0} \left[ \frac{\mu_{0i}^A \mu_{i0}^B}{\Delta_i - \omega} + \frac{\mu_{0i}^B \mu_{i0}^A}{\Delta_i + \omega} \right] = \hat{P}[A(-\omega), B(\omega)] \sum_{i \neq 0} \frac{\mu_{0i}^A \mu_{i0}^B}{\Delta_i - \omega} \quad (5.2)$$

$$\beta_{ABC}(-\omega_\sigma; \omega_1, \omega_2) = \hat{P}[A(-\omega_\sigma), B(\omega_1), C(\omega_2)] \sum_{i=0} \sum_{j \neq 0} \frac{\mu_{0i}^A \overline{\mu_{ij}^B} \mu_{j0}^C}{(\Delta_i - \omega_\sigma)(\Delta_j - \omega_2)} \quad (5.3)$$

A, B, C... such labels are used to indicate the direction X, Y, Z. $\omega$ is the energy of the external field, and when it is 0, it corresponds to the static (super) polarizability. Addition is the addition of all excited states, and $\Delta$ represents the excitation energy of the excited state relative to the ground state. P represents all possible permutations of items in square brackets. For example, for $\beta$ and P, there are three items in square brackets, so there are 3!=6 permutations, and the results of these 6 cases must be added. $\mu_{ij}^A$ represents the A-direction component of the transition dipole moment of the two states i and j. When i=j, it corresponds to the dipole moment of the i-th state, so $\mu_{00}$ is the dipole moment of the ground state. $\theta_{ij}$ is the Kronecker symbol, 1 when i=j, and 0 otherwise.

Similarly, the formulas for the second and third hyperpolarizabilities are

$$\gamma_{ABCD}\left(-\omega_\sigma;\omega_1,\omega_2,\omega_3\right) = \hat{P}\left[A\left(-\omega_\sigma\right),B\left(\omega_1\right),C\left(\omega_2\right),D\left(\omega_3\right)\right]\left(\gamma^{\mathrm{I}}-\gamma^{\mathrm{II}}\right)$$

$$\gamma^{\mathrm{I}} = \sum_{i\neq0}\sum_{j\neq0}\sum_{k\neq0}\frac{\mu_{0i}^A\mu_{ij}^B\mu_{jk}^C\mu_{k0}^D}{(\Delta_i-\omega_\sigma)(\Delta_j-\omega_2-\omega_3)(\Delta_k-\omega_3)}$$

$$\gamma^{\mathrm{II}} = \sum_{i\neq0}\sum_{j\neq0}\frac{\mu_{0i}^A\mu_{i0}^B\mu_{0j}^C\mu_{j0}^D}{(\Delta_i-\omega_\sigma)(\Delta_i-\omega_1)(\Delta_j-\omega_3)}$$

(5.4)

$$\delta_{ABCDE}\left(-\omega_\sigma;\omega_1,\omega_2,\omega_3,\omega_4\right) = \hat{P}\left[A\left(-\omega_\sigma\right),B\left(\omega_1\right),C\left(\omega_2\right),D\left(\omega_3\right),E\left(\omega_4\right)\right]$$
$$\left(\delta^{\mathrm{I}}-\delta^{\mathrm{II}}-\delta^{\mathrm{III}}\right)$$

$$\delta^{\mathrm{I}} = \sum_{i,j,k,l}\frac{\mu_{0i}^A\overline{\mu_{ij}^B\mu_{jk}^C\mu_{kl}^D}\mu_{10}^E}{\left(\Delta_i-\omega_\sigma\right)\left(\Delta_j-\omega_\sigma+\omega_1\right)\left(\Delta_k-\omega_3-\omega_4\right)\left(\Delta_l-\omega_4\right)}$$

$$\delta^{\mathrm{II}} = (1/2)\sum_{\substack{i,j,k\\(\neq0)}}\frac{\mu_{0i}^A\mu_{i0}^B\mu_{0j}^C\overline{\mu_{jk}^D}\mu_{k0}^E}{\left(\Delta_j+\omega_2\right)\left(\Delta_k-\omega_4\right)}$$

$$\left(\frac{1}{\Delta_i-\omega_\sigma}+\frac{1}{\Delta_i-\omega_1}\right)\left(\frac{1}{\Delta_j-\omega_3-\omega_4}+\frac{1}{\Delta_k+\omega_2+\omega_3}\right)$$

$$\delta^{\mathrm{IIII}} = (1/2)\sum_{\substack{i,j,k\\(\neq0)}}\frac{\mu_{0i}^A\mu_{i0}^B\mu_{0j}^C\overline{\mu_{jk}^D}\mu_{k0}^E}{\left(\Delta_i-\omega_\sigma\right)\left(\Delta_i-\omega_1\right)}$$

$$\left[\frac{1}{\left(\Delta_j-\omega_3-\omega_4\right)\left(\Delta_k-\omega_4\right)}+\frac{1}{\left(\Delta_j+\omega_2\right)\left(\Delta_k+\omega_2+\omega_3\right)}\right]$$

(5.5)

It can be seen that it is not difficult to do SOS calculations. The formulas are readily available, as long as the excitation energy, dipole moment of each excited state, and the transition dipole moment between each excited state are provided. These quantities can be produced by electronic excitation methods such as ZINDO, CIS, TDHF, and TDDFT. CIS(D) can also be used. When used in SOS, the transition dipole moment is still CIS, but the excitation energy is corrected by second-order perturbation to better consider the electronic correlation effect.

The entire SOS calculation process is divided into two parts (1) electron excitation calculation (2) cyclic accumulation according to the SOS formula. For part (2), the calculation of $\alpha$ and $\beta$ itself takes almost no time, and $\gamma$ takes a little time. For $\theta$, when there are many states to consider, it can be seen from the above formula that a quadruple cycle accumulation of excited states is required, and it contains $3^5 = 243$ components (although some components are the same to avoid repeated calculations), so calculate $\theta$ It's still quite time-consuming. For $\alpha$, $\beta$, and $\gamma$ that we are generally interested in, the entire calculation cost of SOS is mainly used (1), especially for the case of large systems and high-quality basis sets. Note that the time-consuming part of (2) has no direct relationship with the number of basis functions, but only depends on the number of states considered, but the time-consuming of (1) is directly related to the number of basis functions.

ZINDO is a semi-empirical method, and the calculation is fast. The SOS/ZINDO combination is very cheap and is often used to calculate the NLO properties of organic large conjugate systems. In principle, the more accurate CIS/TDHF/TDDFT is obviously much more time-consuming. In principle, SOS needs to sum all states. Although it is not necessary to consider all states in actual research, it is generally necessary to calculate 40–120 states during the electronic excitation process, which is much higher than the number of states required for general research on electronic excitation. The more states to solve, the more time-consuming CIS/TDHF/TDDFT is. Originally this kind of ab initio electronic excitation calculation is more difficult to use in very large systems. In addition, it is necessary to calculate so many polymorphisms for SOS, and it is more accurate to calculate $\beta$, especially $\gamma$, and requires a large dispersion function. The basis set, SOS combined with CIS/TDHF/TDDFT can be applied to a very limited scale. Compared to SOS, if all derivatives can be calculated analytically, it is a better choice to calculate the (hyper)polarizability using the derivative method. However, it is difficult to support high-order analytic derivatives from a programming point of view. Gaussian's methods to achieve third-order analytic derivatives (corresponding to $\beta$) are only HF, DFT, and semi-empirical methods. They do not support fourth-order analytic derivatives to generate $\gamma$. Although the static $\gamma$ can be obtained by doing a finite difference based on the third-order analytic derivative, the frequency-containing $\gamma$ cannot be obtained, so SOS must be used. In addition, as long as the information required for SOS calculation is available, each SOS calculation is fast, just a simple loop and addition, subtraction, multiplication, and division. Therefore, it is convenient to study the changes of $\alpha$, $\beta$, and $\gamma$ with frequency.

## 5.2.2   Calculation of SHG

The SOS function of Multiwfn can easily calculate the polarization rate and the first, second, and third hyperpolarization rate. The calculation efficiency is very high and it is fully parallelized. Moreover, it is very convenient to calculate the changes of $\alpha$, $\beta$, and $\gamma$ with the number of states considered to test the convergence of SOS, and the changes of $\alpha$, $\beta$, and $\gamma$ with the changes of the external field frequency. Therefore, it can be used not only to calculate the SHG coefficient, but also to calculate the relative susceptibility of other nonlinear optical processes. Note that TPA is a third-order nonlinear optical process, but its section is related to the imaginary part of the second hyperpolarizability, so Multiwfn cannot calculate the TPA section for the time being.

When Multiwfn does SOS, the information to be read includes ground state and excited state dipole moments, excitation energy, and transition dipole moments between all states. This information can be read from two channels

(1) Read directly from Gaussian's ZINDO/CIS/TDHF/TDDFT output file. Unfortunately, the output file of the TDHF/TDDFT task does not contain the transition dipole moments between the excited states. Although CIS and ZINDO have the alltransitiondensities keyword to output the transition dipole moments between the excited states, there is no way to do it all at once. The dipole moment of each excited state is given, and this information is needed to calculate the hyperpolarizability. Therefore, if the Gaussian output file is read directly, the information obtained is only enough to calculate the polarization rate, so at this time Multiwfn also only outputs the polarization rate.

(2) Text file. The user can write the information needed for SOS calculations generated by various quantization programs into a text file in a format similar to the following, and then read it when Multiwfn starts and calculate the polarization and hyperpolarization rates.

The following example is a complete calculation of nonlinear optical coefficients through the SOS method, especially the main process of SHG. The first is Gaussian calculation:

```
#p wB97XD/aug-cc-pVTZ td=nstates=150 IOp(9/40=4)
[Blank line]
opted structure
[Blank line]
0 1
C  0.00000000  1.21057933  0.69594119
C  0.00000000  0.00000000  1.38106446
C -0.00000000 -1.21057933  0.69594119
C -0.00000000 -1.21057933 -0.69594119
C  0.00000000  0.00000000 -1.38106446
C  0.00000000  1.21057933 -0.69594119
H -0.00000000  2.14498731  1.24162617
H -0.00000000 -2.14498731  1.24162617
H -0.00000000 -2.14498731 -1.24162617
H  0.00000000  2.14498731 -1.24162617
Cl 0.00000000  0.00000000 -3.13748669
Cl -0.00000000  0.00000000  3.13748669
```

It is worth noting that the first is the number of states calculated. As mentioned earlier, calculating SOS requires counting many excited states. Generally speaking, the lower the excited state, the greater the influence on the SOS result (but individual higher-order excited states may also have a great influence), and the result will gradually converge as the number of states considered increases. In this example, it is sufficient to consider 150 states, and the result must have converged. This point will be specifically tested below. The calculation of (hyper)polarization rate requires a basis set with rich dispersion functions to get better quantitative results. The higher the order of hyperpolarization rate, the higher the requirements for the number of dispersion functions and angular momentum. The basis set here is Aug-cc-pVTZ. It is enough to calculate the first hyperpolarizability. For the purpose of calculating SOS, it is necessary to write IOp $(9/40 = 5)$. Because by default, only MO transition coefficients greater than 0.1 will be output to the output file, and smaller ones will not be output. In this case, Multiwfn calculates the dipole moments of each state based on these coefficients and the result of the transition dipole moments between them Will be inaccurate. The meaning of IOp$(9/40 = x)$ is to output all configurations with coefficients greater than $10^x$, so IOp$(9/40 = 5)$ will output all configurations with absolute coefficients greater than 0.00001. In this case The result given by Multiwfn is accurate enough (x is set to greater than 5 meaningless, because the Gaussian output format has only 5 decimal places).

Then use formchk to convert the calculated chk file into a .fch file. Then start Multiwfn and enter:

```
[Load fchk file]
18 //Electronic excitation analysis function
5 //Calculate the transition dipole moment between states
[Output file path]
3 //Combine the calculated results, together with various SOS
    required information such as excitation energy, into the input
    file required by Multiwfn's SOS function
```

Soon the calculation is completed and the SOS.txt file is generated in the current directory, and its format is the same as that described in the previous section.

```
Restart Multiwfn and enter
SOS.txt
200 //Main function 200 (collection of miscellaneous functions)
8 //Calculate (super) polarizability using SOS method
```

# Chapter 6
# Calculation and Analysis of Molecular Chiral Spectra

## 6.1 Chirality

The term chirality means that an object cannot coincide with its mirror image. Like our hands, the left hand does not overlap with the right hand that mirrors each other. The term chirality is more commonly used in the field of chemistry and medicine. A chiral molecule and its mirror image do not overlap [26]. The chirality of a molecule is usually caused by asymmetric carbon, that is, the four groups on a carbon are different from each other. It is usually identified by (RS) and (DL). Chiral phenomena also exist widely in nature. Chirality is the basic attribute of nature. Chirality widely exists in nature and represents an important symmetry characteristic in many disciplines. If an object is different from its mirror image, it is called "chiral", and its mirror image cannot be superimposed with the original object, just as the left hand and right hand are mirror images of each other and cannot be superimposed. Chiral objects and their mirror images are called enantiomorphs ("relative/opposite form" in Greek); they are also called enantiomers in references to molecular concepts. Objects that can be superimposed with its mirror image are called achiral, sometimes also called amphichiral. The term optical activity is used to explain the interaction between chiral materials and polarized light. A solution of chiral molecules can rotate the plane of polarized light. This phenomenon was discovered by Jean Baptiste Biot in 1815 and has shown importance in the sugar industry, analytical chemistry, and pharmaceutical fields. Louis Pasteur speculated in 1848 that the chiral phenomenon originated from molecules. In 1961, thalidomide (thalidomide) was fully recalled because of its strong teratogenic effect. Further studies have shown that the R-configuration molecule of thalidomide has a curative effect, while the S-configuration molecule has a strong teratogenic effect. The thalidomide incident has made the chirality of the drug widely valued by the pharmaceutical industry. In 2001, William Standish Knowles, Ryoji Noyori and Barry Sharpless shared the Nobel Prize in chemistry for their contributions to chiral catalysis.

© Tsinghua University Press 2023

M. Sun and X. Mu, *Computational Simulation in Nanophotonics and Spectroscopy*, Nanoscience and Nanotechnology, https://doi.org/10.1007/978-981-99-4732-4_6

Before chiral drugs were not recognized, some doctors in Europe used to give pregnant women racemic drugs as analgesics or cough suppressants. Many pregnant women gave birth to children with congenital deformities without heads or legs. Some fetuses don't have arms, their hands are on their shoulders, and they look terrifying. In just 4 years, more than 12,000 deformed "seal babies" were born worldwide. This is the tragedy called "Thylan". Later, after research, it was found that the R body of thalidomide has a sedative effect, but the S-enantiomer has a strong teratogenic effect on embryos. It is with this lesson from the 1960s that after the drug is successfully developed, it must undergo strict biological activity and toxicity tests to avoid the harm of another chiral molecule contained in it to the human body.

In chemical synthesis, these two molecules appear in equal proportions, so for pharmaceutical companies, for every kilogram of drug they produce, they have to work hard to separate the other half. If there is no use value for them, they can only be waste. In the era of increasingly stringent environmental protection regulations, these waste products cannot be disposed of at will. Considering the possible harm to public health, the disposal of these industrial wastes is also a considerable expense. Therefore, the pharmaceutical company is eager to find a way to solve this problem. For example, if he wants a left-handed molecule, then he has to find a way to convert the other half of the right-handed molecule into a left-handed molecule. Today, this headache has been resolved. Scientists use a method called "asymmetric catalytic synthesis" to solve this problem. This method can be widely used in chemical industries such as pharmaceuticals, flavors and sweeteners, and it has brought huge benefits to industrial production at once. This research also won the Nobel Prize in Chemistry in 2001. There is no doubt that this result is of great significance. Competitive research, while creating industrial miracles, has also inspired us to re-understand the origin of life on earth and even the origin of the universe.

We know that in all aspects of nature, especially in physics and chemistry, there are many symmetrical concepts widely: negatively charged electrons and positively charged counterelectrons, the south and north poles of the magnetic field, and the decomposition and chemical reaction. Even distant galaxies outside the galactic system also have positive and counter-rotating vortex structures. Scientists can't help but wonder: Does this remind us that there is a peculiar universal symmetry law in the universe? There is no D-amino acid life on the earth, but according to the principle of chirality, they are indeed possible, and even intelligent D-amino acid life also exists.

Natural products are important molecules with ecological defense functions produced during the survival and evolution of organisms. Clarifying the structure and function of natural products can reveal the nature of coexistence and reproduction of different species in nature from the molecular level, and can characterize the genetic characteristics and biological classification of species. At the same time, natural products are also important molecular sources of innovative drugs. The structural diversity of natural products endows its biological activity diversity and uniqueness. It is an important task for natural product chemists to excavate secondary metabolites from terrestrial and marine sources and clarify their structural diversity, novelty and biological functions. The stereochemical changes of natural products directly affect their biological activity And targeting, so clarifying the stereochemical structure of natural products is the main content of natural product structure research. With the development of stereochemistry research methods and laws, new technologies and methods for solving the structure of new natural products, especially stereochemistry, are constantly emerging.

Since the 1980s, the rapid development of modern spectroscopy has provided important tools for the structural analysis of complex natural products. Early stereochemistry analysis of natural products mostly used chemical methods, including chemical degradation and chiral synthesis. These techniques require more samples and are time-consuming and labor-intensive. With the deepening of natural product chemistry research, chemists pay more attention to the excavation of micro-molecules in organisms, which play an important role in the signal transduction and biological information transmission processes in organisms. The development of micro-NMR technology provides the possibility to analyze the structure of nanogram-level trace natural molecules. The invention of single crystal X-ray diffraction technology provides a convenient way to analyze the three-dimensional configuration of natural products that can cultivate single crystals. However, it is difficult for a considerable number of natural products to form crystals, especially for the crystallization of trace molecules. The three-dimensional configuration of these compounds is difficult to solve with single crystal X-ray diffraction technology. Among the commonly used spectroscopic techniques, modern nuclear magnetic resonance technology has played a huge role in the structure analysis of natural products, but the nuclear magnetic resonance methods used for configuration determination, such as NOE detection technology NOESY, ROESY, GOESY and coupling constant determination, mainly provide Information about the relative configuration of molecular structures. At present, a variety of non-single crystal diffraction methods, including specific rotation, optical rotation spectroscopy (ORD), circular dichroism (CD), Morsher method and Marfey method, have been applied to determine the absolute configuration of natural products. In recent years, with the development of computer technology, the calculation of CD (ECD and VCD) methods have also been widely used to determine the absolute configuration of chiral natural products.

## 6.2   Chiral Spectroscopy

### 6.2.1   Electron Circular Dichroism (ECD)

Determining the absolute configuration of chiral compounds is very important in organic chemistry. The most rigorous method is X-ray single crystal diffraction method to solve the precise structure of the compound, and then the absolute configuration can be known. However, it is not easy to obtain single crystals in experiments, so other spectroscopy methods are often needed to make judgments. Circular dichroism (CD) is one of the commonly used methods. The difference in optical properties of chiral enantiomers is mainly manifested in the response to polarized light. When the left circularly polarized light and the right circularly polarized light pass through the chiral compound solution, the propagation rate and absorption degree of the left and right circularly polarized light will change. The circular dichroism can be obtained by plotting the difference in molar absorption coefficient ($\Delta\epsilon$) with wavelength. If the system has no chirality, there is no CD signal. There are two types of circular dichroism: electronic circular dichroism (ECD) and vibrational circular dichroism (VCD). In ECD spectroscopy, the absorption of plane-polarized light by chiral compounds is caused by the transition between electron energy levels produced by electrons absorbing photons, and belongs to electronic absorption spectra. The corresponding absorption spectrum of VCD is vibrational spectrum, which is produced by transitions between different vibrational energy levels in the same electronic energy state. This article introduces the calculation method of electronic circular dichroism. Experimental chemists generally directly refer to circular dichroism as electronic circular dichroism.

ECD belongs to electronic absorption spectroscopy. For theoretical calculation of ECD, only conventional excited state calculations are required. The most commonly used method is TD-DFT. In other words, the conventional TD-DFT calculation will give the UV–Vis spectrum and the ECD spectrum at the same time. Unlike the ultraviolet spectrum, the difference $\Delta\epsilon_k$ between the left and right polarization absorption coefficients corresponding to the transition from the ground state to the kth excited state is determined by the rotation intensity $R_k$ corresponding to the transition. The ECD spectra between the isomers that are mirror images of each other are completely symmetrical (positive and negative), so only one configuration can be calculated to compare with the experimental results and determine the conformation of the compound.

The ECD spectrum is very sensitive to the structure, and different conformations of the same configuration can even give very different spectra. Therefore, it is necessary to weight the spectra of different conformations according to their probability distribution to obtain the final spectrum. But in the calculation of quantum chemistry, for example, the calculation using Gaussian software is to calculate the speed rotation intensity of the conformation (Velocity rotatory strengths, R), for example [27]:

$$R = 2.296 \times 10^{-39} \int \frac{\Delta\varepsilon(v)}{v} dv \qquad (6.1)$$

Whether the absolute configuration of a chiral substance can be identified using ECD depends on the basic structure of the molecule. It requires that the chiral center of the chiral molecule should be closer to the UV chromophore group, such as C=C, C=O, benzene ring, etc., usually not more than 3 atoms away, preferably in the ortho position or In between. When the chromophore is on the ring and the chiral center is in the chain structure, if the chiral center is next to the chromophore, the absolute configuration of the molecule can also be determined by the ECD method. From the point of view of electronic transition, the ECD intensity can be expressed as:

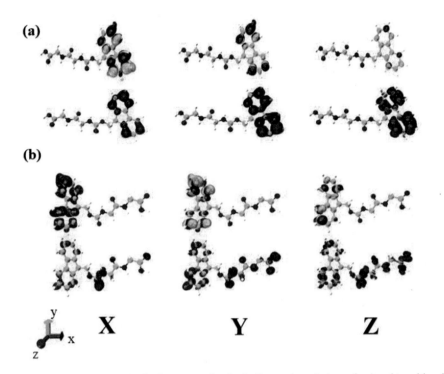

**Fig. 6.1** Transition electric dipole moment density (yellow and purple isosurface) and transitional magnetic dipole moment density (red and blue isosurface) component isosurface map of chiral molecular

$$I \propto \left|\langle\varphi_j \,|\mu_e|\, \varphi_i\rangle E\right|^2 + \left|\langle\varphi_j \,|\mu_e|\, \varphi_i\rangle\langle\varphi_j \,|\mu_m|\, \varphi_i\rangle B\right|^2 \qquad (6.2)$$

Therefore, the ECD intensity and the transition electric/magnetic dipole moment are both related. Therefore, visualizing these two densities is very important for analyzing the mechanism of ECD spectrum generation, that is, the chiral mechanism. It can be seen from Fig. 6.1 that the transition electric dipole moment and the transition magnetic dipole moment are distributed in different places. These regions are closely related to molecular chirality [28].

In fact, the function of this analysis method is not only in the mechanism analysis of traditional molecular chirality. Since this visualization method is not limited to the system, it can analyze the group chiral light response of the cluster system and the material system. The cluster dynamics is shown in Fig. 6.2. This assembly molecular cluster has the chiral optical properties and the physical mechanism is visualized by transition electric dipole moments in Fig. 6.3 [29]. In addition, this method with same physical theory can also applied in the twisted bilayer graphene (TwBLG). With this visualization method, the chiral of TwBLG is proved by the transition electric dipole moment, which distributed in the moire pattern, see Fig. 6.4 [30].

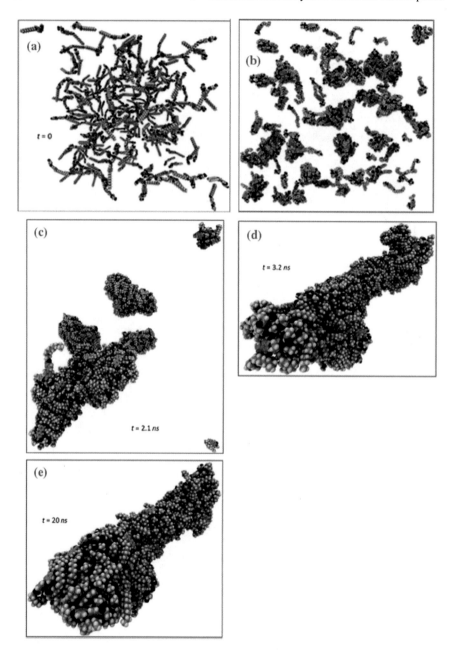

**Fig. 6.2** The molecular assembly dynamics at different time scale in (**a**)–(**e**)

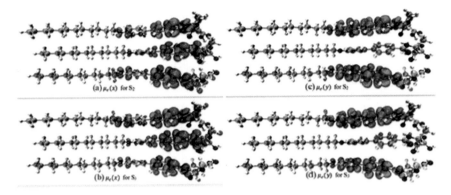

**Fig. 6.3** The transition electric dipole moments density isosurface of assembly molecular cluster

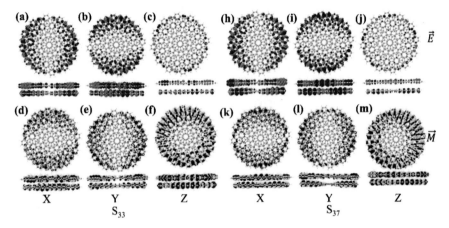

**Fig. 6.4** The transition electric/magnetic dipole moment of different excited states in TwBLG GQDs

## 6.2.2 Raman Optical Activity (ROA)

Raman optical activity (ROA) is a kind of vibrational spectroscopy that can determine the chirality of a compound based on the difference in intensity between left-handed and right-handed polarized light of Raman scattered light. This instrument was originally studied by Laurence D Barron and Peter Atkins at Oxford University [31, 32]. Then it was further developed by Barron and A. David Buckingham at Cambridge University. Then many related developments were completed by Werner Hug of the University of Freiburg and Laurence Barron of the University of Glasgow, such as instruments related to the practical application of Raman optical activity. To put it simply, Raman rotation is based on the interaction between polarized light and the optical rotation tensor of the anti-palm isomers, resulting in different intensity of left-handed and right-handed polarized light of Raman scattering. The difference of the spectral intensity of different wave numbers can be used to determine the relevant information of the three-dimensional center in the sample. The form of Raman rotation depends on the polarization of incident light and scattered light. For example, in the experiment of scattered circular polarization (SCP), the difference between scattered light and linearly polarized incident light is measured; another example is in dual circular polarization (DCP) Among them, both the incident light and the scattered light are circularly polarized light, and the main measurement is DCPI or DCPII (Phase difference).

Gaussian software can calculate ROA spectra. Since ROA spectrum is the chiral response of the analysis system to different wavelengths of light, it is necessary to specify the excitation wavelength. The CPHF keyword can be used in Gaussian to specify the excitation wavelength after the program reads the molecular coordinates. In fact, the number combination unit (nm) specifies the excitation wavelength, and the value without a unit is the incident photon energy defined in the atomic unit of energy (Hartree). The program example is as follows:

```
b3lyp/6-31g* freq=ROA CPHF=Rdfreq
[Blank line]
test
[Blank line]
0 1
C -0.00000000 0.00000000 0.00000000
H -0.00000000 0.00000000 1.09000000
H -0.00000000 -1.02766186 -0.36333333
H -0.88998127 0.51383093 -0.36333333
H 0.88998127 0.51383093 -0.36333333
[Blank line]
532nm 633nm
```

The definition of ROA intensity based on electronic transition is [27, 28]:

$$
\begin{aligned}
-I \propto \sum_{k \neq ji} &\left| \frac{\langle \varphi_j | \mu_e | \varphi_k \rangle \langle \varphi_k | \mu_e | \varphi_i \rangle}{\omega_{ji} - \omega_0 + i\Gamma} \right|^2 E^4 \\
+ &\left| \sum_{i \neq j} \frac{\langle \varphi_j | \mu_e | \varphi_i \rangle \langle \varphi_j | \mu_m | \varphi_i \rangle}{\omega_{ji} - \omega_0 + i\Gamma} + \sum_{i \neq j} \frac{\langle \varphi_j | \theta_e | \varphi_i \rangle \langle \varphi_j | \mu_e | \varphi_i \rangle}{\omega_{ji} - \omega_0 + i\Gamma} \right|^2 E^4
\end{aligned}
\tag{6.3}
$$

It can be found from the second term of the formula that the transition electrical quadrupole moment ($\langle \varphi_j | \theta_e | \varphi_i \rangle$) is the main factor causing ROA. Therefore, following the previous analysis method of ECD mechanism, the electric quadrupole moment can also be visualized. It can be found that the ROA spectra of retinal molecules show

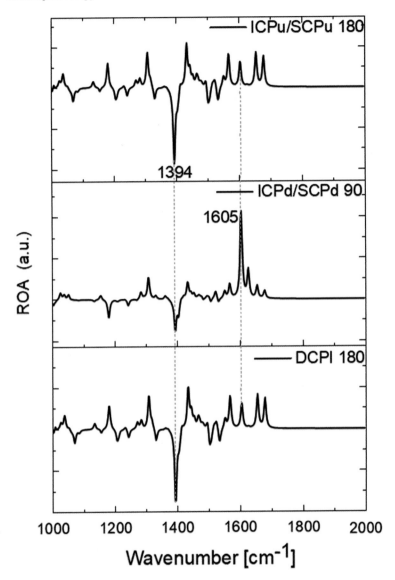

**Fig. 6.5**  The ROA spectra of PSB11 excited with 4.512 eV, which is the same energy of S3 electronic transition of PSB11

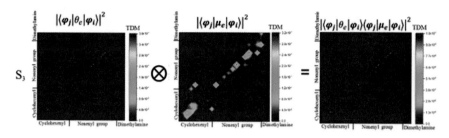

**Fig. 6.6**   The 2D visualization color map of transition quadrupole moment in PSB11

different vibrational chirality at different frequencies, see Fig. 6.5. Through the visual analysis of the transition electric quadrupole moment, the distribution of different chiral ROA responses in the molecule can be found, see Fig. 6.6.

# Chapter 7
# First Principles Calculation of Optical Properties of Solids

## 7.1 Optical Properties of Solids

When light passes through a solid, light absorption occurs due to the interaction of light with electrons, excitons, lattice vibrations and defects in the solid. When the solid absorbs external energy, part of the energy is emitted in the form of light. The photoelectric phenomena of solids include: light absorption, photoconductivity, photovoltaic effect and light emission. The optical properties of a solid, in essence, are the interaction between the solid and electromagnetic waves, which involves the reflection and absorption of light radiation by the crystal, the luminescence of the crystal under the action of light, the propagation and function of light in the crystal, and the photoelectric effect, light Magnetic effect and so on. Based on these properties, optical crystal materials, optoelectronic materials, luminescent materials, laser materials, and various optical function conversion materials can be developed. The energy band structure of electrons in a solid, in which the valence band is equivalent to the valence electron layer of an anion, is completely filled with electrons. There is a certain width of energy gap (forbidden band) between the conduction band and the valence band, and the energy level of electrons cannot exist in the energy gap. In this way, when the solid is exposed to light, if the energy of the radiated photon is not enough to make the electron transition from the valence band to the conduction band, then the crystal will not be excited, nor will it absorb light. For example, the energy gap width of ionic crystals is generally several electron volts, which is equivalent to the energy of ultraviolet light. Therefore, pure ideal ion crystals will not absorb light from visible light and even infrared radiation, and are transparent. The transparent wavelength of the alkali metal halide crystals to electromagnetic waves can range from 25 μm to 250 nm, which is equivalent to an energy of 0.05–5 eV. When there is sufficiently strong radiation (such as violet light) to irradiate the ion crystal, electrons in the valence band may be excited to cross the energy gap and enter the conduction band, thus light absorption occurs. This light absorption related to the transition of electrons from the valence band to the conduction band is called

© Tsinghua University Press 2023

M. Sun and X. Mu, *Computational Simulation in Nanophotonics and Spectroscopy*,
Nanoscience and Nanotechnology,
https://doi.org/10.1007/978-981-99-4732-4_7

basal absorption or intrinsic absorption. For example, the basic absorption band of CaF2 is around 200 nm (about 6 eV), the basic absorption of NaCl is about 8 eV, and the basic absorption of Al2O3 is about 9 eV.

In addition to the basic absorption, there is also a type of absorption whose energy is lower than the energy gap width, which corresponds to the transition of electrons from the valence band to the energy level slightly below the bottom of the conduction band. These energy levels can be regarded as the excitation energy levels of some electron-holes (or called exciton, excition).

## 7.2  Light Absorption of Inorganic Solids

The band gap of inorganic ion solids is relatively large, generally a few electron volts, which is equivalent to the energy in the ultraviolet region. Therefore, when visible light or even infrared light irradiates the crystal, such energy is not enough to cause electrons to cross the energy gap and transition from the valence band to the conduction band. Therefore, the crystal will not be excited, nor will light absorption occur, and the crystals are transparent. When ultraviolet light irradiates the crystal, light absorption occurs and the crystal becomes opaque. The relationship between the band gap $E_g$ and the absorption wavelength $\lambda$ is

$$E_g = hv = hc/\lambda$$
$$\lambda = hc/E_g \tag{7.1}$$

However, after the impurity ions are introduced into the inorganic ion crystal, an electron-hole recombination process will occur between the impurity defect level and the valence band level, and the corresponding energy will be smaller than the interband width E.g., which often falls in the visible light region. As a result, light absorption by the solid occurs. For example, $Al^{3+}$ and $O^{2-}$ ions in Al2O3 crystals are combined in a hexagonal close-packed manner under the action of electrostatic attraction. The ground state energy level of $Al^{3+}$ and $O^{2-}$ ions is a closed electron shell filled with electrons, and its energy gap is 9ev. It cannot absorb visible light, so it is transparent. If 0.1% of $Cr^{3+}$ is added to it, the crystal will be pink. When 1% of $Cr^{3+}$ is added, the crystal will be dark red. This is ruby, which can absorb visible light and emit fluorescence. This is because the doped $Cr^{3+}$ ions have a shell filled with electrons, causing a part of the lower excited state energy level in the Al2O3 crystal, which can absorb visible light. In fact, this material is a typical laser material.

Crystal defects are intrinsic, such as interstitial atoms and vacancies, as well as extrinsic, such as substitute impurities. The energy level of these defects is set in the energy gap between the valence band and the conduction band. When the material is exposed to light, the acceptor defect energy level accepts electrons that migrate from the valence band, and the electrons on the donor energy level can migrate to the conduction band, so that the substance that cannot have basic absorption will absorb light due to the existence of defects.

## 7.3 Optical Properties of Semiconductor

### 7.3.1 Intrinsic Semiconductor Light Absorption

The electronic energy band structure of an intrinsic semiconductor is similar to that of an insulator. All electrons are filled in the valence band and are fully filled, but there are no electrons in the conduction band, but the energy gap between the valence band and the conduction band is small, about 1 eV. At extremely low temperatures, all electrons are in the valence band and will not move in any direction. It is an insulator, and its optical properties are the same as the aforementioned insulators. When the temperature rises, some electrons may gain sufficient energy to cross the energy gap and jump into the originally empty conduction band. At this time, empty energy levels appear in the valence band and electrons appear in the conduction band. If an electric field is applied, conduction will occur. Therefore, the band gap of the semiconductor material at room temperature determines the material's properties. The light absorption and luminescence of intrinsic semiconductors generally originate from the transition of electrons across the energy gap, that is, direct transition. The electrons in the valence band absorb a certain wavelength of visible light or near-infrared light, can separate from each other and drift by themselves, and participate in conduction, which is the so-called photoconductivity phenomenon. When an electron in the conduction band recombines with a hole in the valence band, a photon of visible light is emitted, which is the so-called photoluminescence phenomenon.

### 7.3.2 Extrinsic Semiconductor Light Absorption

There are three types of doped semiconductor impurities: donor impurities, acceptor impurities, and isoelectronic impurities. The energy levels of these impurities are localized in the energy gap, which constitutes the various light absorption transition modes shown in Fig. 1.3. The presence of isoelectronic impurities may become the center of the recombination of electrons and holes, which will affect the luminescence of the material, and the individual donor and acceptor impurities will not affect the optical properties of the material. This is because only when the excited state electrons cross the energy gap and recombine with the holes, the semiconductor will emit light. For example, an n-type semiconductor can provide enough electrons in the conduction band, but there are no holes in the valence band, so it will not emit light. Similarly, a p-type semiconductor has holes in the valence band, but there are no electrons in its conduction band, so it does not emit light. If the n-type semiconductor and the p-type semiconductor are combined to form a p-n junction, the excited state electrons (from the conduction band of the n-type semiconductor) and holes (from the valence band of the p-type semiconductor) can be recombined at the p-n junction. We apply a forward bias voltage at the p-n junction, which can inject the conduction band electrons in the n region into the valence band in the p region, where they recombine with holes to generate photon radiation. In the basic absorption of light by the crystal, electrons and holes become carriers at the same time, which contribute to the conductance of the crystal. In the impurity absorption of the crystal, the electrons excited into the conduction band can participate in conduction, but the remaining holes are bound in the impurity center and cannot participate in conduction. Such holes capture adjacent electrons and recombine. When the valence band electrons are excited to the impurity center by light, the holes generated in the valence band can participate in conduction. The carriers generated by the excitation of light radiation, on the one hand, disappear at the load center, on the other hand, they can move a certain distance under the action of an electric field before being trapped. If the strength of the external electric field is large, the carriers will drift in the crystal for a long distance before being captured by the trap, and the photocurrent will be strong, but there will be a saturation value (that is, the maximum value of the primary photocurrent).

## 7.4   Calculation of Solid Optical Properties in Common Software

The famous first-principles software Vienna Ab initio Simulation Package (VASP) can easily calculate the optical properties of solids. Of course, some commercial software can also calculate the optical properties, and the principles used are exactly the same as VASP. These methods first calculate the imaginary part of the dielectric function by calculating the transition probability between the bands, and firstly calculate the imaginary part of the dielectric function according to the probability of the transition between the bands and the imaginary part of the dielectric function, which is defined as follows:

$$\epsilon_{\alpha\beta}^{(2)}(\omega) = \frac{4\pi^2 e^2}{\Omega} \lim_{\zeta \to 0} \frac{1}{q^2} \sum_{c,v,\mathbf{k}} 2w_{\mathbf{k}} \delta \left( \epsilon_{c\mathbf{k}} - \epsilon_{u\mathbf{k}} - \omega \right) \times \left\langle u_{c\mathbf{k}+\mathbf{e}_\alpha q} \mid u_{v\mathbf{k}} \right\rangle \left\langle u_{v\mathbf{k}} \mid u_{c\mathbf{k}+\mathbf{e}_\beta q} \right\rangle$$

$$(7.2)$$

where the $\left\langle u_{c\mathbf{k}+\mathbf{e}_\alpha q} \mid u_{v\mathbf{k}} \right\rangle \left\langle u_{v\mathbf{k}} \mid u_{c\mathbf{k}+\mathbf{e}_\rho q} \right\rangle$ is the inter-band transition probability. Then by analyzing the continuous nature of the complex variable function in the upper half plane, the real part of the dielectric function is calculated by the following formula.

$$\epsilon_{\alpha\beta}^{(1)}(\omega) = 1 + \frac{2}{\pi} P \int_0^\infty \frac{\epsilon_{\alpha\beta}^{(2)}(\omega')\, \omega'}{\omega^2 - \omega^2 + i\eta} d\omega'$$

$$(7.3)$$

where the $P$ is the principle value. The dielectric function is the basis of all optical properties. The frequency-dependent dielectric function can be used to obtain the refractive index, absorption coefficient, extinction spectrum, energy-loss function, and reflectivity coefficients by simple transformation.

$$n(\omega) = \left( \frac{\sqrt{\varepsilon_1^2 + \varepsilon_2^2} + \varepsilon_1}{2} \right)^{\frac{1}{2}}$$

$$k(\omega) = \left( \frac{\sqrt{\varepsilon_1^2 + \varepsilon_2^2} - \varepsilon_1}{2} \right)^{\frac{1}{2}}$$

$$\alpha(\omega) = \sqrt{2}\omega \left( \frac{\sqrt{\varepsilon_1^2 + \varepsilon_2^2} - \varepsilon_1}{2} \right)^{\frac{1}{2}} \tag{7.4}$$

$$L(\omega) = \mathrm{Im}\left( \frac{-1}{\varepsilon(\omega)} \right) = \frac{\varepsilon_2}{\varepsilon_1^2 + \varepsilon_2^2}$$

$$R(\omega) = \frac{(n-1)^2 + k^2}{(n+1)^2 + k^2}$$

In fact, the calculation of optical properties in VASP software, that is, frequency-dependent optical properties, is very simple. After structural optimization and static calculation (for pre-calculation please refer to: https://www.vasp.at/wiki/index.php/The_VASP_Manual), you can use the following keywords to calculate optical properties.

```
ALGO = Exact
NBANDS  = 64
LOPTICS = .TRUE. ; CSHIFT = 0.100
NEDOS = 2000

## and you might try with the following
#LPEAD = .TRUE.

ISMEAR =  0
SIGMA  =  0.01
EDIFF  = 1.E-8
```

Afterwards, data can be extracted through the following Shell script. Of course, there are also many convenient auxiliary programs to complete this step, and other optical properties can be transformed for mapping and post research. For example, the VASPKIT program developed by Wang et al. (http://vaspkit.sourceforge.net/).

```
awk 'BEGIN{i=1} /imag/,\
/\/imag/ \
{a[i]=$2 ; b[i]=$3 ; i=i+1} \
END{for (j=12;j<i-3;j++) print a[j],b[j]}' vasprun.xml > imag.dat

awk 'BEGIN{i=1} /real/,\
/\/real/ \
{a[i]=$2 ; b[i]=$3 ; i=i+1} \
END{for (j=12;j<i-3;j++) print a[j],b[j]}' vasprun.xml > real.dat

cat >plotfile<<!
# set term postscript enhanced eps colour lw 2 "Helvetica" 20
# set output "optics.eps"
plot [0:25] "imag.dat" using (\$1):(\$2) w lp, "real.dat" using (\$1):(\$2) w lp
!

gnuplot -persist plotfile
```

Of course the above example is only suitable for pure functional calculation of optical properties. As we all know, pure functionals have the disadvantage of underestimating the band gap. For this shortcoming, hybrid functionals (HSE06, etc.) can be used to solve this problem. Of course, VASP also has a built-in method that is more accurate in certain situations, that is, the Bethe–Salpeter equation (BSE). The advantage of this method is that the exciton effect in the solid can be considered. The specific calculation is divided into many steps. First, the ground state wave function needs to be obtained:

```
System  = Si
PREC = Normal ; ENCUT = 250.0
ALGO = EXACT ; NELM = 1
ISMEAR = 0 ; SIGMA = 0.01
KPAR = 2
NBANDS = 128
LOPTICS = .TRUE. ; LPEAD = .TRUE.
OMEGAMAX = 40
```

This step is also to get the virtual track. Then perform GW calculation:

```
System  = Si
PREC = Normal ; ENCUT = 250.0
ALGO = GW0
ISMEAR = 0 ; SIGMA = 0.01
ENCUTGW = 150 ; NELM = 1 ;  NOMEGA =  50 ;   OMEGATL = 280
KPAR = 2
#NBANDSO=4 ; NBANDSV=8 ; LADDER=.TRUE. ; LUSEW=.TRUE.
NBANDS = 128
NBANDSGW = 12
LWAVE = .TRUE.
PRECFOCK = Normal
```

This step is to obtain the energy of the vertically excited quasiparticle as follows:

```
System  = Si
PREC = Normal ; ENCUT = 250.0
ALGO = Nothing ; NELM = 1
ISMEAR = 0 ; SIGMA = 0.01
KPAR = 2
NBANDS = 128
LWAVE = .FALSE.
LOPTICS = .TRUE. ; LPEAD = .TRUE.
OMEGAMAX = 40
```

Then BSE calculation can be performed:

```
PREC = Normal ; ENCUT = 250.0
ALGO = BSE
ANTIRES = 0
ISMEAR = 0 ; SIGMA = 0.01
ENCUTGW = 150
EDIFF = 1.E-8
```

```
NBANDS = 128
NBANDSO = 4
NBANDSV = 8
OMEGAMAX = 20
PRECFOCK = Normal
```

Please pay attention to the different algorithms in different steps (ALGO key-words). Especially the BSE algorithm in the last step.

## 7.5 Application of Solid Optical Properties in Surface Plasmon

Using the above-mentioned method to calculate the optical properties of the van der Waals constant structure and the lateral in-plane heterostructure, it can be found that both heterostructures in the visible region and the near-infrared region have light absorption. In addition, by analyzing the weak peaks of electron-hole pairs in the near-infrared region, significant charge transfer can be found, see Fig. 7.1.

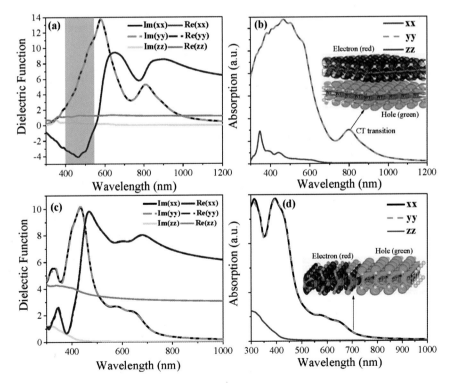

**Fig. 7.1** The dielectric function and absorption of vdW (**a**, **b**) and lateral (**c**, **d**) heterostructure, respectively [33]

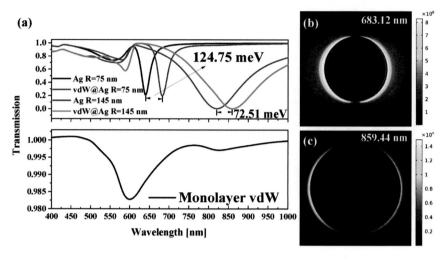

**Fig. 7.2** The transmission spectra and surface plasmom electric mode of Ag disk and vdW heterostructure, respectively

Secondly, it can be found that the optical properties in the xx and yy directions completely coincide. This is because the material is a uniaxial crystal, and its optical property matrix is defined as follows:

$$s = \begin{bmatrix} \varepsilon_{xx} & & \\ & \varepsilon_{yy} & \\ & & \varepsilon_{zz} \end{bmatrix} \tag{7.5}$$

The above matrix satisfies the following relationship: $\varepsilon_{xx} = \varepsilon_{yy} \neq \varepsilon_{zz}$.

Finally, the real part of the dielectric function of the green region in Fig. 7.1a is less than zero and the imaginary part is greater than zero. It shows that the materials in this region show good plasmon properties. In other words, the plasmonic properties of materials such as different structures and impurities can be studied through first principles. This is a very important method for subwavelength optics, nanophotonics and plasmon optics. Because the dielectric function calculated by the first principles can be well applied in the finite element program, as an input parameter for the plasmon study of different materials, as shown in Fig. 7.2 [33].

# Chapter 8
# Application of Electronic Structure Method in Optical Calculation and Analysis

## 8.1 Energy Band Theory

Energy band theory is a theory that uses quantum mechanics to study the movement of electrons inside a solid. It is an approximate theory developed after the establishment of quantum mechanics in the early 20th century. It has qualitatively explained the general characteristics of electron movement in crystals, and further explained the difference between conductors, insulators and semiconductors, and explained the problem of the mean free path of electrons in crystals.

Since electronic computers were widely used in the 1960s, it has become possible to use electronic computers to calculate complex energy band structures based on first principles. The energy band theory has developed from qualitative to a quantitative and accurate science. The energy band structure of solid materials consists of multiple energy bands, similar to the energy levels of electrons in atoms. The electron first occupies the low-energy band, and gradually occupies the high-energy band. According to the electronic filling, the energy band is divided into conduction band (referred to as conduction band, small amount of electron filling) and valence band (referred to as valence band, large amount of electron filling). The gap between the conduction band and the valence band is called the forbidden band (electrons cannot be filled), and the size is the energy gap.

The energy band structure can explain the origin of the three major differences between conductor (no energy gap), semiconductor (energy gap < 3 eV), and insulator (energy gap > 3 eV) in solids. The conductivity of a material is determined by the number of electrons contained in the "conduction band". When electrons obtain energy from the "valence band" and jump to the "conduction band", under the action of an external electric field, the electrons in the unfilled conduction band energy band generate a net current, and the material exhibits conductivity.

Generally common metal materials, because the "energy gap" between the conduction band and the valence band is very small, electrons can easily gain energy at room temperature and jump to the conduction band to conduct electricity, while

© Tsinghua University Press 2023
M. Sun and X. Mu, *Computational Simulation in Nanophotonics and Spectroscopy*,
Nanoscience and Nanotechnology,
https://doi.org/10.1007/978-981-99-4732-4_8

insulating materials have a large energy gap (usually more than 3 eV), it is difficult for electrons to jump to the conduction band, so they cannot conduct electricity. Generally, the energy gap of semiconductor materials is about 1–3 eV, which is between conductors and insulators. Therefore, the material can conduct electricity as long as it is given appropriate conditions of energy excitation, or changes the distance between its energy gaps. The energy band theory believes that the electrons inside a solid are not bound around a single atom, but move inside the entire solid, only being perturbed by the real potential field of the ion. The main part of the eigen-wave function is the eigenstate of momentum, and scattering only gives a first-order correction. This model is mainly suitable for metals.

Energy band theory is currently the most important theoretical basis for studying the state of electrons in solids. Its appearance is the most direct and important result of the application of quantum mechanics and quantum statistics in solids. The energy band theory successfully solved many problems left over by Sommerfeld's free electron theory when dealing with metal problems, and laid the foundation for the development of solid-state physics.

The basic starting point of energy band theory is that the electrons in a solid are no longer completely bound around an atom, but can move in the entire solid, which is called a shared electron. But the electron is not like free electrons in the process of movement, which is completely free from any force. The electron is affected by the potential field of the lattice atom during the movement.

## 8.1.1 Fundamental Assumption

In statistical physics and solid-state physics, when discussing the lattice Brillouin zone, it is assumed that the atoms in the lattice are stationary in the equilibrium position. In fact, the atoms in the crystal undergo thermal vibration. This will have a certain impact on the movement of electrons. Since the mass of the nucleus is much greater than the mass of the electron, its movement speed is much slower than that of the electron.

Since the mass of the electron is much smaller than the mass of the nucleus, the movement of the electron and the nucleus can be treated separately, that is, only the Coulomb effect of the nucleus on the electron is considered, and the other two effects are not considered, which is equivalent to the nucleus providing only external potential to the electron. .

Under the BO approximation, the multi-body system formed by electrons and atomic nuclei of the crystal is transformed into the classical mechanical motion of the atomic nucleus on the crystal lattice and the quantum mechanical motion of multiple electrons. The motion of the atomic nucleus is approximate to simple harmonic motion, which can be regarded as many The linear superposition of lattice waves, the quantum of lattice waves is phonons; and the interacting multi-electron system is described by Schrodinger equation.

## 8.1.2 Conduction Band

The conduction band is the energy space formed by free electrons. For semiconductors, the energy band of all valence electrons is the so-called valence band, and the energy band with higher energy than the valence band is the conduction band. At absolute zero temperature, the valence band of a semiconductor is full. After photoelectric injection or thermal excitation, part of the electrons in the valence band will cross the forbidden band and enter the empty band with higher energy. The existence of electrons in the empty band becomes conductive Energy band-conduction band.

Donor and acceptor: For doped semiconductors, most of the electrons and holes are provided by impurities. Impurities that can donate electrons are called donors; impurities that can donate holes are called acceptors. The energy level of the donor is in the forbidden band near the bottom of the conduction band; the energy level of the acceptor is in the forbidden band near the top of the valence band. In fact, the Fermi level of an undoped semiconductor is near the center of the valence band and conduction band. The Fermi level of the n-type semiconductor is near the bottom of the conduction band, while the p-type is near the top of the valence band.

Potential energy/kinetic energy: The bottom of the conduction band is the lowest energy level of the conduction band, which can be regarded as the potential energy of electrons. Usually, the electron is near the bottom of the conduction band; the height of energy leaving the bottom of the conduction band can be regarded as the kinetic energy of the electron. When an external field acts on both ends of the semiconductor, the potential energy of the electron changes, and the band diagram shows that the bottom of the conduction band is tilted; conversely, wherever the band is tilted, there must be an electric field (external electric field).

## 8.1.3 Valence and Forbidden Band

The valence band usually refers to the highest energy band in a semiconductor or insulator that can be filled by electrons at 0 K. For semiconductors, the energy levels in this energy band are basically continuous. The electrons in the full energy band cannot move freely in the solid. But if the electron is exposed to light, it can absorb enough energy to jump into the next allowable highest energy zone, so that the valence band becomes partially filled, and the electrons left in the valence band can move freely in the solid.

Band gap is often used to indicate the energy interval between the valence band and the conduction band where the density of energy states is zero. The size of the band gap determines whether the material has semiconductor properties or insulator properties. Semiconductors have a small forbidden band width. When the temperature rises, electrons can be excited to pass to the conduction band, thus making the material conductive. The forbidden band width of the insulator is very large, even at higher temperatures, it is still a poor conductor of electricity. The band gap of inorganic semiconductors ranges from 0.1 to 2.0 eV, the band gap of $\pi - \pi$ conjugated polymers is approximately 1.4–4.2 eV, and the band gap of insulators is greater than 4.5 eV.

"Electron concentration = hole concentration" is actually a characteristic of intrinsic semiconductors. Therefore, it can be said that any semiconductor with the same carrier concentration is an intrinsic semiconductor. Note: Not only undoped semiconductors are intrinsic semiconductors, but doped semiconductors can also be transformed into intrinsic semiconductors under certain conditions (such as high temperatures). Holes, carriers: Many electrons (valence electrons) in the valence band cannot conduct electricity, but a small amount of valence electron vacancies—holes can conduct electricity, so holes are called carriers. The lowest energy of the hole—potential energy, that is, the top of the valence band, usually the hole is near the top of the valence band. Band gap: The energy difference between the top of the valence band and the bottom of the conduction band is the so-called band gap of the semiconductor. This is the minimum average energy required to generate intrinsic excitation. Energy gap (energy band gap, in solid state physics refers to the energy gap between the top of the valence band of a semiconductor or insulator to the bottom of the conduction band.

## 8.2  Density of States (DOS)

In statistical mechanics and condensed matter physics, the density of states or density of states is the number of microscopic states per unit energy interval near a certain energy, also called the density of states. In physics, a microscopic state with the same energy is called degenerate. The number of degenerate states is called the degenerate number. At discrete energy levels, the degenerate number is the density of states of the corresponding energy. In continuous and quasi-continuous energy states, the $g(E)$ density of states, then the number of states in the energy E and E+dE interval is $g(E)\mathrm{d}E$. The importance of the density of states lies in the probability is $\rho(E)\mathrm{d}E \propto g(E)\exp(-\beta E)\mathrm{d}E$ that the system is between energy E and E+dE in a canonical ensemble, where is Boltzmann's constant. Considering normalization,

$$\rho(E) = \frac{g(E)\exp(-\beta E)}{\int_0^\infty g(E)\exp(-\beta E)\mathrm{d}E} \tag{8.1}$$

In fact, the system partition function can be used to express DOS in statistical physics.

$$Z(\beta) = \int_0^\infty g(E)\exp(-\beta E)\mathrm{d}E \tag{8.2}$$

The partition function in the above formula can be transformed into DOS after Laplace transform.

$$g(E) = \frac{1}{2\pi i} \int_{s-i\infty}^{s+i\infty} e^{\beta E} Z(\beta) \mathrm{d}\beta \quad (\Re s > 0) \tag{8.3}$$

For electrons, Fermi gas can be used to approximate its behavior, then the DOS of Fermi gas can be expressed as:

$$g(E) = \frac{gV}{h^3} 4\pi p^2 \frac{\partial p}{\partial E}\bigg|_E \tag{8.4}$$

Among them, g is the number of inner degrees of freedom of the fermion (such as spin, quark, etc.), and V is the volume. The relationship between momentum p and energy E is called dispersion relationship. The dispersion relation of non-relativistic fermions is $E = \frac{p^2}{2m}$. Therefore, the density of state of the non-relativistic zero-temperature ideal Fermi gas is,

$$g(E) = \frac{g(2m)^{3/2} V}{4\pi^2 \hbar^3} \sqrt{E} \tag{8.5}$$

Similarly, the dispersion relation of extreme relativistic fermions is $E = pc$. Therefore, the relativistic zero-temperature ideal Fermi gas has the density of state as

$$g(E) = \frac{gV}{2\pi^2 \hbar^3 c^3} E^2. \tag{8.6}$$

## 8.3 Effective Mass

Effective mass (Effective mass) is an approximation of Newton's second law used to facilitate the introduction of classical mechanics. It is approximately considered that the electron is subjected to the periodic potential field of the nucleus (this potential field is the same as the period of the lattice) and the result of the combined effect of other electronic potential fields. In the mathematical processing, Taylor expansion is used at the extreme points of the energy band, so that the terms above the second order can be omitted, and it can be well described. In addition, the effective mass is related to the shape and position of the energy band. Including the effective mass of state density and the effective mass of conductance. Effective mass is a very important concept, which directly links the acceleration of the quasi-classical movement of electrons in the crystal with external forces.

The second-order approximation of the effective mass is:

$$E(\mathbf{k}) = E_0 + \frac{\hbar^2 \mathbf{k}^2}{2m^*} \tag{8.7}$$

The physical properties of most semiconductors are mainly determined by the carriers near the extreme point of the energy band. That is to say: (1) Only the band structure near the extreme point of the band is important; (2) Quantitative physics should come from the iso-energy surface. These properties can be obtained through symmetry analysis. For the first point, to study the movement of carriers, we need a relatively accurate band structure near the bottom of the conduction band and the top of the valence band. Although the tight-binding approximation gives the band structure in the whole space, it is relatively inaccurate near the local k point, and other methods cannot solve this problem well. At the same time, none of these methods can get a quantitative E-k relationship, and can only give an approximate situation in the vicinity. The traditional E-k method, like the work of Dresslhaus and Kane, solves this problem, and the physical image is very clear. According to the symmetry of the k-point, a reasonable Hamiltonian is constructed to give the energy band structure of the semiconductor material near this point, and then calculate the remaining physical quantities. In this process, a series of physical parameters related to the experiment were introduced, and the results obtained were compared with the experiment. In fact, the k-p method far away from the extreme point has also been given by Cardona and Pollak. They used the full-space theory to give the real electronic structures of Si and Ge. Therefore, it is an empirical calculation method of energy band structure. For the second point, it shows the subtlety of symmetry analysis, just like the invariant method introduced by Luttinger. On the other hand, with the development of semiconductor process technology, various microstructure quantum wells, quantum wires, and quantum dots have been manufactured. On the basis of the bulk material k-p model, an envelope function expansion method has been developed to deal with such nanostructures. This method has had a profound impact on semiconductor process devices. The physical parameters proposed by the model were extracted and applied to the semiconductor device model, which developed the computer-aided software TCAD for various simulation devices. This kind of software is based on the k-p theory and uses computer simulation technology to simulate electronic devices [34]. Most of the physics of k-p theory can be demonstrated through a single-band model. Such a single band can be the conduction band of a semiconductor or sometimes the valence band. We will demonstrate how to derive the k-p Hamiltonian of semiconductor materials. The k-p Hamiltonian is obtained by the Schrodinger equation of a single electron. Bloch wave function and cell wave function satisfy the following orthogonality relationship:

$$
\begin{aligned}
\langle \psi_{nk} \mid \psi_{n'k'} \rangle &= \int dV \psi_{nk}^*(\mathbf{r}) \psi_{n'k'}(\mathbf{r}) = \delta_{n,n'} \delta\left(\mathbf{k} - \mathbf{k}'\right) \\
\langle u_{nk} \mid u_{n'k'} \rangle &= \int d\Omega u_{nk}^*(\mathbf{r}) u_{n'k'}(\mathbf{r}) = \delta_{n,n'} \frac{\Omega}{(2\pi)^3}
\end{aligned}
\tag{8.8}
$$

Since carriers only occupy the lowest conduction band and the highest valence band, the relative importance of remote energy bands is decreasing. For cubic semiconductors, the most important energy band is the conduction band. The effective mass of the conduction band is isotropic:

$$\frac{1}{m_e} = \frac{1}{m_0} + \frac{2}{m_0^2} \frac{|\langle S | p_x | X_v \rangle|^2}{E_{\Gamma_{1c}} - E_{\Gamma_{15v}}} + \frac{2}{m_0^2} \frac{|\langle S | p_x | X_c \rangle|^2}{E_{\Gamma_{1c}} - E_{\Gamma_{15c}}}$$

$$\equiv \frac{1}{m_0} + \frac{2P^2}{\hbar^2 E_0} - \frac{2P'^2}{\hbar^2 E_0'} \tag{8.9}$$

among them

$$P^2 = \frac{\hbar^2}{m_0^2} |\langle S | p_x | X_v \rangle|^2$$

$$P'^2 = \frac{\hbar^2}{m_0^2} |\langle S | p_x | X_c \rangle|^2 . \tag{8.10}$$

## 8.4  Application of Electronic Structure Method

If the energy gap is small or 0, the solid is a metal material. At room temperature, electrons can easily gain energy and jump to rewind and conduct electricity; while insulating materials have a large energy gap (usually greater than 9 eV), electrons It is difficult to jump to the conduction band, so it cannot conduct electricity. Generally, the energy gap of semiconductor materials is about 1–3 eV, which is between conductors and insulators. Therefore, as long as the energy excitation is given to the appropriate conditions, or the distance between the energy gaps is changed, the material can conduct electricity. The energy band is used to qualitatively clarify the general characteristics of electron movement in the crystal. Valence band, or valence band, usually refers to the highest energy of electrons in a solid material at absolute zero. In the conduction band, the energy range of electrons is higher than the valence band, and all electrons in the conduction band can be accelerated by an external electric field to form a current. For semiconductors and insulators, there is an energy above the valence band. Band gap, the energy band above the band is the conduction band. Only after electrons enter the conduction band can they move freely in the solid material to form an electric current. For metals, there is no band gap between the valence band and the conduction band. Therefore, the valence band refers specifically to the condition of semiconductors and insulators. The fermi level is the highest energy level at absolute zero. According to Pauli's exclusion principle, a quantum state cannot contain two or more Fermions (electrons), so at absolute zero, electrons will fill the energy levels from low to high, and all but the highest energy level will be filled, forming a "Fermi sea" of electronic states. "Fermi sea" "The average energy of each electron is (under absolute zero) $\frac{3}{5}$ of the Fermi level. Sea level is the Fermi level. Generally speaking, the Fermi level corresponds to the place where the density of states is 0, but For insulators, the Fermi level is at the top of the valence band. The prerequisite for becoming a good electronic conductor is that the Fermi level intersects one or more energy bands. Therefore, the semiconductor Fermi level has energy at 0 DOS. Dispersion. The reason why there are quantum states with different energies in the same energy band is that the electrons in the energy band have different wave vectors, or k-vectors. In quantum mechanics, k-vectors are the momentum of particles, and different Materials will have different energy-momentum relationships). Energy dispersion determines whether the energy gap of a semiconductor material is a direct energy gap or an indirect energy gap. If the K value of the lowest point of the conduction band is the same as the highest point of the valence band, it is a direct energy gap, otherwise it is an indirect energy gap. The width of the energy band. The width or three degrees of the energy band, that is, the energy difference between the highest and lowest energy levels of the energy band, is a very important feature. It is determined by the overlap between the interacting orbitals, which reflects the difference between the orbitals. The overlap between adjacent tracks, the greater the overlap between adjacent tracks, the greater the bandwidth. The abscissa of the energy band diagram is the K point taken on the basis of the symmetry of the model. Why take K points? Because of the periodicity of the crystal, the solution of Schrodinger's equation is also periodic. Taking K points according to symmetry can ensure that the most complete energy feature solution is obtained with the smallest amount of calculation. The abscissa of the energy band diagram is K point, which is actually a geometric point in the inverted space. The ordinate is energy. Then the energy band diagram should represent the energy of each point with symmetry in the research system. The total energy of the system we get should be the sum of the energy of all points in the entire system. The qualitative/quantitative discussion is mainly from the following three aspects:

1. Charge density diagram;
2. Energy band structure;
3. Density of states (DOS).

The charge density map appears in the article in the form of a graph, which is very intuitive, so there will be no doubts for the average entry-level researcher. The only thing that needs to be noted is the various derivative forms of this analysis, such as differential charge density maps and quadratic differential maps, etc. Spin polarization may also include spin polarization charge density maps. The so-called "differential" refers to the redistribution of charge after the atom is composed of a system (cluster), and "secondary" refers to the redistribution of the charge after the chemical composition or geometric configuration of the same system is changed, so this difference diagram can be very intuitive Observe the bonding of atoms in the system. According to the specific spatial distribution of charge accumulation/loss, it can be regarded as the polarity of the bond; the orbital of the bond is judged by the shape of the charge distribution near a certain lattice point (this is mainly the analysis of the d orbital, for the shape of the s or p orbital I have not seen the analysis). The method of analyzing the total charge density map is similar, but relatively speaking, the amount of information carried by this map is relatively small. The charge density difference of charge transfer before and after bonding.

Band structure analysis is now very common in first-principles calculations in various fields. First of all, of course, it can be seen whether this system is a metal, a semiconductor or an insulator. For intrinsic semiconductors, it can also be seen whether it is a direct energy gap or an indirect energy gap: if the lowest point of the conduction band and the highest point of the valence band are at the same k point, it is a direct energy gap, otherwise it is an indirect energy gap.

(1) Because most of the current calculations are in the form of superunit cells, there are dozens of atoms and hundreds of electrons in a unit cell, so the band diagrams obtained are often very flat far below the Fermi level. Very dense. In principle, the energy band in this area does not have much interpretation/reading value. Therefore, don't be frightened by this phenomenon. In general work, we are mainly concerned with the band shape near the Fermi level.

(2) The width of the energy band occupies a very important position in the analysis of the energy band. The wider the energy band, that is, the greater the fluctuations in the energy band diagram, it means that the smaller the effective mass of the electrons in this band, the greater the degree of non-localization, and the stronger the expansibility of the atomic orbitals that make up this energy band. If the shape is similar to a parabolic shape, it will generally be named as the sp-like zone. Conversely, a relatively narrow energy band indicates that the eigenstate corresponding to this energy band is mainly composed of atomic orbitals localized at a certain lattice point. The electrons on this band are very localized, and the effective mass is relatively high.

(3) If the system is a doped extrinsic semiconductor, pay attention to the comparison with the energy band structure diagram of the intrinsic semiconductor. Generally speaking, a new, narrower energy band will appear at the energy gap. This is the so-called impurity state, or acceptor state or donor state according to the type of doped semiconductor.

(4) Regarding the energy band of spin polarization, two pictures are generally drawn: majority spin and minority spin. Classically speaking, it represents the energy band structure composed of spin-up and spin-down orbitals. Note the difference between them at the Fermi level. If the Fermi level intersects the energy band diagram of the majority spin and is in the energy gap of the minority spin, the system has obvious spin polarization, and the system can also be called a semi-metal. If the energy band where the majority spin intersects the Fermi level is mainly composed of impurity atomic orbitals, this can be the starting point to discuss the magnetic characteristics of the impurity.

(5) When dealing with interface problems, the energy band diagram of the substrate material is very important, and there may be different situations between the high symmetry points. Specifically, between certain two points, the Fermi level intersects the energy band; while in another interval of k, the Fermi level is exactly between the conduction band and the valence band. In this way, the substrate material exhibits anisotropy: for the former, it is metallic, and for the latter, it is insulating. Therefore, some work is to select a different surface as the growth surface through the band diagram of a certain material. The specific analysis should be given in conjunction with the test results.

In principle, the density of states can be used as a visual result of the band structure. Many analysis and energy band analysis results can be one-to-one correspondence, and many terms are also similar to energy band analysis. But because it is more intuitive, it is more widely used than energy band analysis in the discussion of results. A brief summary of the main points of analysis is as follows:

(1) DOS, which is relatively evenly distributed and has no local spikes in the entire energy range, corresponds to the sp-like band (this statement needs to be verified-bloggers), indicating the non-localized nature of electronics Very strong. On the contrary, for general transition metals, the DOS of the d orbital is generally a large spike, indicating that the d electron is relatively localized and the corresponding energy band is relatively narrow.

(2) The energy gap characteristics can also be analyzed from the DOS diagram: if the Fermi level is in the range where the DOS value is zero, it means that the system is a semiconductor or an insulator; if there is a partial wave DOS that crosses the Fermi level, the system is metal. In addition, two kinds of density of states (PDOS) and local area (LDOS) can be drawn, and the bonding of the partial waves at each point can be studied in more detail.

(3) The concept of "pseudogap" (pseudogap) can also be introduced from the DOS diagram. That is, there are two spikes on both sides of the Fermi level. The DOS between the two spikes is not zero. The pseudo-energy gap directly reflects the strength of the covalent bond of the system: the wider, the stronger the covalent. If the analysis is the local density of states (LDOS), the pseudo-energy gap reflects the strength of the bond between two adjacent atoms: the wider the pseudo-energy gap, the stronger the bond between the two atoms. The theoretical basis of the above analysis can be explained from the tight-binding theory: in fact, it can be considered that the width of the pseudo-energy gap is directly related to the non-diagonal elements of the Hamiltonian matrix, and they are in a monotonically increasing functional relationship with each other.

(4) For a spin-polarized system, similar to energy band analysis, major spin and minor spin should also be drawn separately. If the Fermi level intersects with the major DOS and is within the energy gap of the minor DOS, you can Explain the spin polarization of the system.

(5) Consider LDOS. If the LDOS of adjacent atoms has a spike at the same energy at the same time, we call it a hybrid peak. This concept intuitively shows us the strength of the interaction between adjacent atoms. Because the energy band of the metal may cross the fermi energy level, it causes discontinuous changes in the total energy calculation. In order to avoid this situation, it is necessary to introduce fractional occupied state smearing.

Through the theoretical analysis of electronic structure, phenomena in nanophotonics can be connected with electronic structure, and the internal physical mechanism of optical phenomena can be explained. Figure 8.1 shows the energy bands, density of states and anisotropic effective mass of two different heterostructures of MoS-WS2. The energy band binding state density can reveal the intrinsic physical mechanism of

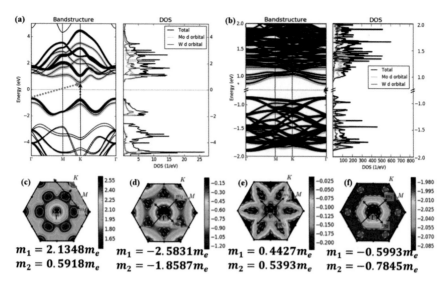

**Fig. 8.1** The band structure, DOS and effective mass of vdW and lateral heterostructure [33]

the spectrum in Fig. 7.1. The effective mass can be seen as strong anisotropy, so the electromagnetic field mode in the interaction of surface plasmons (Fig. 7.2) is related to anisotropy (the transport phenomenon of different mobility of hot electrons).

# References

1. Alec Michael Hammond. Machine learning methods for nanophotonic design, simulation, and operation. 2019.
2. IS Komarov and EA Bezus. Development of a software package for modeling and analysis of light diffraction on periodic structures of nanophotonics by the rigorous coupled-wave analysis. In *Journal of Physics: Conference Series*, volume 1368, page 022058. IOP Publishing, 2019.
3. Constantine Sideris, Emmanuel Garza, and Oscar P Bruno. Ultrafast simulation and optimization of nanophotonic devices with integral equation methods. *ACS Photonics*, 6(12):3233–3240, 2019.
4. Tyler W Hughes, Momchil Minkov, Ian AD Williamson, and Shanhui Fan. Adjoint method and inverse design for nonlinear nanophotonic devices. *ACS Photonics*, 5(12):4781–4787, 2018.
5. NL Kazanskiy and PG Serafimovich. Cloud computing for rigorous coupled-wave analysis. *Advances in Optical Technologies*, 2012, 2012.
6. Sharon C Glotzer, Peter Nordlander, and Laura E Fernandez. Theory, simulation, and computation in nanoscience and nanotechnology, 2017.
7. John Peurifoy, Yichen Shen, Li Jing, Yi Yang, Fidel Cano-Renteria, Brendan G DeLacy, John D Joannopoulos, Max Tegmark, and Marin Soljačić. Nanophotonic particle simulation and inverse design using artificial neural networks. *Science advances*, 4(6):eaar4206, 2018.
8. Nikolay L Kazanskiy and Pavel G Serafimovich. Cloud computing for nanophotonic simulations. In *International Workshop on Optical Supercomputing*, pages 54–67. Springer, 2012.
9. Arthur J Freeman. Materials by design and the exciting role of quantum computation/simulation. *Journal of computational and applied mathematics*, 149(1):27–56, 2002.
10. Anubhav Jain, Shyue Ping Ong, Geoffroy Hautier, Wei Chen, William Davidson Richards, Stephen Dacek, Shreyas Cholia, Dan Gunter, David Skinner, Gerbrand Ceder, et al. Commentary: The materials project: A materials genome approach to accelerating materials innovation. *Apl Materials*, 1(1):011002, 2013.
11. ACT North, DC t Phillips, and F Scott Mathews. A semi-empirical method of absorption correction. *Acta Crystallographica Section A: Crystal Physics, Diffraction, Theoretical and General Crystallography*, 24(3):351–359, 1968.
12. Rudolph Pariser and Robert G Parr. A semi-empirical theory of the electronic spectra and electronic structure of complex unsaturated molecules. i. *The Journal of Chemical Physics*, 21(3):466–471, 1953.

© Tsinghua University Press 2023

M. Sun and X. Mu, *Computational Simulation in Nanophotonics and Spectroscopy*,

Nanoscience and Nanotechnology,

https://doi.org/10.1007/978-981-99-4732-4

13. Frank Jensen. Polarization consistent basis sets. iii. the importance of diffuse functions. *The Journal of chemical physics*, 117(20):9234–9240, 2002.
14. Catherine E Check, Timothy O Faust, John M Bailey, Brian J Wright, Thomas M Gilbert, and Lee S Sunderlin. Addition of polarization and diffuse functions to the lanl2dz basis set for p-block elements. *The Journal of Physical Chemistry A*, 105(34):8111–8116, 2001.
15. Ewa Papajak and Donald G Truhlar. Efficient diffuse basis sets for density functional theory. *Journal of chemical theory and computation*, 6(3):597–601, 2010.
16. F Di Meo, Patrick Trouillas, Carlo Adamo, and Juan-Carlos Sancho-Garcia. Application of recent double-hybrid density functionals to low-lying singlet-singlet excitation energies of large organic compounds. *The Journal of chemical physics*, 139(16):164104, 2013.
17. Zeyu Liu, Tian Lu, and Qinxue Chen. An sp-hybridized all-carboatomic ring, cyclo [18] carbon: Electronic structure, electronic spectrum, and optical nonlinearity. *Carbon*, 2020.
18. Tangui Le Bahers, Carlo Adamo, and Ilaria Ciofini. A qualitative index of spatial extent in charge-transfer excitations. *Journal of chemical theory and computation*, 7(8):2498–2506, 2011.
19. Tian Lu and Feiwu Chen. Multiwfn: a multifunctional wavefunction analyzer. *Journal of computational chemistry*, 33(5):580–592, 2012.
20. Maria Goppert-Mayer. Uber elementarakte mit zwei quantensprungen. *Ann. Phys.*, 9:273–295, 1931.
21. Xijiao Mu, Jingang Wang, and Mengtao Sun. Visualization of photoinduced charge transfer and electron–hole coherence in two-photon absorption. *The Journal of Physical Chemistry C*, 123(23):14132–14143, 2019.
22. Winfried Denk, James H Strickler, and Watt W Webb. Two-photon laser scanning fluorescence microscopy. *Science*, 248(4951):73–76, 1990.
23. Xijiao Mu, Huan Zong, Lilong Zhu, and Mengtao Sun. External electric field-dependent photoinduced charge transfer in a donor–acceptor system in two-photon absorption. *The Journal of Physical Chemistry C*, 124(4):2319–2332, 2020.
24. Xijiao Mu, Xinxin Wang, Jun Quan, and Mengtao Sun. Photoinduced charge transfer in donor-bridge-acceptor in one-and two-photon absorption: Sequential and superexchange mechanisms. *The Journal of Physical Chemistry C*, 124(9):4968–4981, 2020.
25. Kotoku Sasagane, Fumihiko Aiga, and Reikichi Itoh. Higher-order response theory based on the quasienergy derivatives: The derivation of the frequency-dependent polarizabilities and hyperpolarizabilities. *The Journal of chemical physics*, 99(5):3738–3778, 1993.
26. Robert S Cahn, Christopher Ingold, and Vladimir Prelog. Specification of molecular chirality. *Angewandte Chemie International Edition in English*, 5(4):385–415, 1966.
27. John C Lindon, George E Tranter, and David Koppenaal. *Encyclopedia of spectroscopy and spectrometry*. Academic Press, 2016.
28. Xijiao Mu, Xiangtao Chen, Jingang Wang, and Mengtao Sun. Visualizations of electric and magnetic interactions in electronic circular dichroism and raman optical activity. *The Journal of Physical Chemistry A*, 123(37):8071–8081, 2019.
29. Xijiao Mu, Jingang Wang, Guoqiang Duan, Zhijuan Li, Juxiu Wen, and Mengtao Sun. The nature of chirality induced by molecular aggregation and self-assembly. *Spectrochimica Acta Part A: Molecular and Biomolecular Spectroscopy*, 212:188–198, 2019.
30. Xijiao Mu and Mengtao Sun. The linear and nonlinear optical absorption and asymmetrical electromagnetic interaction in chiral twisted bilayer graphene with hybrid edges. *Materials Today Physics*, page 100222, 2020.
31. PW Atkins and Laurence David Barron. Quantum field theory of optical birefringence phenomena i. linear and nonlinear optical rotation. *Proceedings of the Royal Society of London. Series A. Mathematical and Physical Sciences*, 304(1478):303–317, 1968.

32. LD Barron, MP Bogaard, and AD Buckingham. Raman scattering of circularly polarized light by optically active molecules. *Journal of the American Chemical Society*, 95(2):603–605, 1973.
33. Xijiao Mu and Mengtao Sun. Interfacial charge transfer exciton enhanced by plasmon in 2d in-plane lateral and van der waals heterostructures. *Applied Physics Letters*, 117(9):091601, 2020.
34. Gennadiĭ Levikovich Bir and Grigoriĭ Ezekielevich Pikus. *Symmetry and strain-induced effects in semiconductors*, volume 484. Wiley New York, 1974.

Printed in the United States
by Baker & Taylor Publisher Services